SpringerBriefs in Education

More information about this series at http://www.springer.com/series/8914

Anjum Halai · Geoff Tennant
Editors

Mathematics Education in East Africa

Towards Harmonization and Enhancement of Education Quality

ICMI International Commission on Mathematical Instruction

Springer

Editors
Anjum Halai
Institute for Educational Development
East Africa
Aga Khan University
Dar es Salaam
Tanzania

Geoff Tennant
Institute for Educational Development
East Africa
Aga Khan University
Dar es Salaam
Tanzania

ISSN 2211-1921 ISSN 2211-193X (electronic)
SpringerBriefs in Education
ISBN 978-3-319-27257-3 ISBN 978-3-319-27258-0 (eBook)
DOI 10.1007/978-3-319-27258-0

Library of Congress Control Number: 2015958539

Preface

The International Commission on Mathematical Instruction (ICMI) has a substantive interest in ensuring that its resources and influence extend to any country that is able to mobilize the elements of its mathematics education enterprise to make productive use of this connection. In line with this goal, ICMI in conjunction with the International Mathematical Union (IMU), and with the support of UNESCO and ICSU (International Council for Science), promotes the Capacity & Networking Project (CANP). It aims to enhance mathematics education in developing countries by supporting the educational capacity of those responsible for mathematics teachers, and to create sustained regional networks of teachers, mathematics educators and mathematicians, linking them to international support.

CANP consists of a programme in a different developing world region every year: each programme has, at its centre, a two-week workshop of about forty participants, half from the host country and half from regional neighbours, who interact with experts in mathematics, mathematics education, and school policy coming from different parts of the world. It is primarily aimed at mathematics teacher educators, but each event includes also mathematicians, researchers, policy makers, and key teachers.

The Capacity & Networking Project is a major international initiative in the mathematical sciences in the developing world to help exchange information, share state of the art research, enhance mathematics education and build a sustainable network for policy makers, scholars and practitioners across those targeted regions.

The programme builds on existing activities in the region and does not seek to reproduce or compete with existing development programmes.

At the time when this book is printed (2016) five CANP workshops have been held: CANP-1 in Sub Saharan Africa (2012), CANP-2 in Central America and Caribbean Area (2012), CANP-3 in South East Asia (2013), CANP-4 in East Africa (2014), CANP-5 in Andean Region and Paraguay (2016).

The main goal of a CANP consists in building capacity in mathematics education and creating a sustainable regional network in the countries, which participate in the workshop, with a common goal of improving mathematics education in the region. The initial two-week workshop is an occasion for launching the network and for collecting and sharing information about the situation of mathematics teaching in the region. For this, before the workshop each group of participants from a country prepares a report about the state of the art in their own country: the reports are presented, compared and discussed during the meeting. After that, they are further elaborated according the results of the discussions and constitute a final report for that CANP.

They constitute interesting documents about mathematics education in the regions touched by the different CANPs, and give a piece of information not always accessible in an easy way. For this reason ICMI decided to launch a new series of books with an international publisher, Springer, in order to make accessible non-expensive format reports to an international audience of informed policy makers and scientists.

The present volume is the first in the series of CANP reports: it is the result of a huge work of elaboration of the original documents presented at CANP-4 workshop, held from September 1st to 12th, 2014 at the Aga Khan University Institute for Educational Development East Africa, in Dar es Salaam, Tanzania. The event involved more than 80 participants from Tanzania, Kenya, Uganda and Rwanda; it was organised in a splendid way thanks to the wonderful work both of the International Programme Committee, and of the Local Organising Committee (LOC), and especially of Anjum Halai, chair of the LOC, and her team. The Aga Khan University Institute for Educational Development East Africa supported the event in a number of ways which made it possible to offer such a rich programme. In fact CANP-4 included lectures given by outstanding mathematicians and mathematics educators, regional presentations, workshops, round table discussions, panel presentations, and other parallel activities (school visits, mathematical games, traditional dance, poems). Many hours were devoted to the discussion of the regional reports, which are the germs from which this book was originated, and to the creation of an East Africa Mathematics Education and Research Network, chaired by Alphonse Uworwabayeho from Rwanda who is ably supported by Angelina Bijura from Tanzania.

The editors of the volume, Anjum Halai and Geoff Tennant, and the other authors, Peter Kajoro, Simon Karuku, Mussa Mohamed, Veronica Sarungi, and Alphonse Uworwabayeho, made a huge effort to have the different articles written according to what they call the 'Harmonization and Enhancement of Education Quality' issues. They are nicely illustrated in the book: the main idea is to show the necessity in Eastern Africa countries of equipping the students with those mathematical skills that will enable them to compete effectively in the East African Community's (EAC's) envisaged common market and to facilitate mobility of students and teachers across the EAC partner states. The six chapters of the book show the related difficulties and possibilities for the school systems in EAC, which come from a different colonial past (British and Belgian).

I thank all those who have made possible the existence of this book: the editors, the authors, the publisher, and particularly the participants to the CANP-4 event. I do think that making its content accessible to mathematics educators, teachers, and policy makers represents a useful tool for approaching the problems of mathematics education within a global landscape, but without forgetting the specific cultural and social needs of a developing region, in this case the EAC.

It is my strong hope that with the publication of the other CANP books, we will have a wide updated picture of mathematics education needs and problems from relevant parts of the developing world. This will help to avoid the dangers of the alienation generated by the loss of the variety of cultural richness existing in the different regions of the world.

January 26, 2016

Ferdinando Arzarello
President of the International Commission
on Mathematical Instruction

Contents

Editors and Contributors

About the Editors

Anjum Halai is a professor of education currently working at the Aga Khan University Institute for Educational Development in Tanzania. From teaching mathematics in secondary schools to graduate teaching and supervision, she has substantial experience in the field of mathematics education, especially in technologically less developed countries including Pakistan, Kenya Tanzania and Uganda. Anjum Halai obtained her DPhil from Oxford University in the UK. Her research interests include issues of social justice in classroom settings where learners are marginalized on the basis of language, gender, conflict and other forms of exclusion.

Geoff Tennant is associate professor and head of teaching programmes at the Institute of Educational Development, East Africa, Aga Khan University. He has previously worked in University-based secondary mathematics initial teacher training in the UK. His research interests reflect the concern that all children should have maximal access to the school curriculum, particularly in mathematics, including forming meaningful links between number and introductory algebra.

Contributors

Peter Kajoro is assistant professor at the Aga Khan University Institute for Educational Development, East Africa. He previously taught mathematics and science in schools in Dar es Salaam, Singida and Tabora. He is part of a team writing mathematics textbooks for secondary schools in Tanzania, particularly aimed at supporting students as they make the transition from Swahili to English as the medium of instruction. He is research interests also include teacher education and the use of dynamic geometry software in the mathematics classroom.

Simon Karuku is a lecturer in mathematics education at Embu University College in Kenya. For the past four years, he has taught and researched on mathematics and mathematics education at various institutions, including the Institute for Educational Development, East Africa Aga Khan University. He obtained his PhD in Mathematics Education from the University of Alberta, using a phenomenological approach to analyse the nature of help in the mathematics classroom. His research interests include the dynamics of student–teacher interactions in mathematics teaching and learning.

Mussa Mohamed is a lecturer in science education at the Aga Khan University Institute for Educational Development, East Africa. He has an undergraduate degree in science with education from the University of Dar es Salaam and a Master of Education degree from Aga Khan University. His research interests focus on promoting scientific and mathematical literacy in the early years.

Veronica Sarungi is a lecturer currently working at the Aga Khan University's Institute for Educational Development, East Africa (IED-EA) in Tanzania. She previously worked as a secondary mathematics teacher and is currently involved in mathematics education, including in-service programs for teachers in Tanzania, Kenya and Uganda. Veronica has a Masters in Education specialising in mathematics and science from IED-EA. Her research interests include mathematics teacher education, specific learning difficulties in mathematics and numeracy for early years' learners.

Alphonse Uworwabayeho is senior lecturer at the University of Rwanda College of Education. He was previously a mathematics teacher in secondary schools. He is currently leading the university department responsible for Teacher Training Colleges. He gained his PhD on integrating ICT in the teaching and learning of mathematics at the University of Bristol, United Kingdom. His research interests lie in professional development in the Rwandan context via a model of participatory action research that involves mathematics teachers in developing innovative classroom scenarios through hands on experimentation with ICT and reflection and discussion with other teachers.

Introduction

In the increasingly global and technological world, mathematics is seen as a significant gatekeeper of opportunities for social and economic advancement and mobility. Hence, in the context of a post-2015 scenario (2015 was the target year to achieve the Millennium Development Goals led by UNESCO, see unesco.org), countries and development agencies in the East Africa region and more broadly are looking towards increasing access to a relevant and high-quality secondary education as a lever towards economic development (World Bank 2011; UNESCO 2012). Policy makers and other key decision makers in education are looking towards improvement in mathematics teaching and learning as a key focus in education reform. In the East Africa region also, a number of initiatives have been taken at the national levels to improve the quality of mathematics education. For example, these include the SMASSE[1] initiative in Kenya, the SESEMAT[2] in Uganda and similar initiatives in Tanzania all significantly supporting the quality of mathematics education and/or mathematics teacher education in the respective country.

However, relatively little is known about the quality of *secondary mathematics education* in the East Africa region from the perspective of: (a) mathematics curriculum and syllabus in public secondary schools; (b) teaching and learning in public secondary mathematics classrooms; (c) achievement in mathematics; and (d) mathematics teacher education. Hence, this report aims to present, and discuss critically, the content, process and outcomes of secondary mathematics education in the region, provide a comparative perspective of the issues in mathematics

[1]"Strengthening Science and Mathematics in Secondary Education" (SMASSE), a significant initiative in science and mathematics education, was launched by the Ministry of Education in collaboration with the Japan International Cooperation Agency in 1998 and further extended.

[2]Secondary Science Education and Mathematics Teachers (SESEMAT) Project initiated by the Ministry of Education and Sport in collaboration with the Japan International Cooperation Agency since 2004.

education in the four countries and make recommendations for policy and practice. It addresses the following two broad questions:

- What is the state of secondary mathematics education in East Africa?
- Comparatively, what are the possibilities, issues and challenges in improving the state of secondary mathematics education in the region?

This edited collection is expected to be an invaluable source of information and knowledge for academics, practitioners and policy makers to reflect on their own practices as it brings insights mainly from developing countries where relatively less research activity takes place.

The book is expected also to be a valuable resource for courses in mathematics education and related social sciences both at the graduate and undergraduate levels, as well as for students of international development.

This work was undertaken over a nine-month period (November 1, 2013–July 31, 2014) by a team of experienced mathematics education researchers in the region. Methodology involved looking at: (a) key recent or ongoing initiatives and reform projects in secondary mathematics education undertaken by the respective Ministry of Education independently or in collaboration with international development partners; (b) examination achievement results of the national examination councils in the four countries; (c) syllabus and curriculum content and related documents in the public secondary schools in the four countries; (d) teacher education curriculum and syllabus and related documents; and (e) a review of classroom-based studies and projects in the four countries.

Every effort was made including two cycles of peer review to ensure quality, comprehensiveness, accuracy and validity of information in the book.

References

UNESCO. (2012). *Challenges in basic mathematics education*. Paris: UNESCO. ISBN 978-92-3-00-071-3.

World Bank. (2011). *A regional exploration of the pathways towards harmonization of math & science curriculum in the East African Community*. Ministerial Forum Summary Report.

Chapter 1
Issues for Quality Enhancement and Harmonization of Education in East Africa

Mussa Mohamed, Anjum Halai and Simon Karuku

Abstract This chapter provides a description of the historical and political development of the current mass education system in East Africa. It goes on to provide a comparative analysis of the structure of the school education system in the East African countries, Kenya, Rwanda, Tanzania and Uganda. Finally, on the basis of the historical and comparative analysis of the education system the chapter raises questions for harmonization of the education system in the partner states in the East African Community.

Introduction

The system of formal education as currently found in the East African region has its roots in the colonial history of the region. After independence, the education system was characterized by strong structural continuity but also by significant change. Perhaps owing to their common colonial ancestry, Kenya, Tanzania and Uganda opted for a common education system—the one inherited from the former British administration. On the other hand, Burundi and Rwanda charted their own independent courses of education, closely mirroring the education system of their Belgian colonial masters. In recent years, however, there has been a gradual shift in policy within the East African Community (EAC), informed by the region's

M. Mohamed (✉) · A. Halai
Aga Khan University Institute for Educational Development East Africa,
Salaama House Urambo Street, PO125, Dar es Salaam, Tanzania
e-mail: mussa.mohamed@aku.edu

A. Halai
e-mail: anjum.halai@aku.edu

S. Karuku
Embu University College, P.O. Box 6, Embu 60100, Kenya
e-mail: simon.karuku@embuni.ac.ke

A. Halai and G. Tennant (eds.), *Mathematics Education in East Africa*,
https://doi.org/10.1007/978-3-319-27258-0_1

socio-economic needs. In particular, there is a concerted effort to harmonize the region's education system with the aim of attaining an "increasingly networked and interrelated group of curriculum and examination systems, linked in such a way that these systems possess overlapping, interconnected and comparable logics that are capable of influencing each other across the partner states" (East African Community 2012).

Ultimately, the goal is to equip students with skills that will enable them to compete effectively in the EAC's envisaged common market and to facilitate mobility of students and teachers across the EAC partner states. The catalyst for this harmonization agenda is the ongoing effort to revive the EAC. EAC is a regional intergovernmental organization comprising Burundi, Kenya, Rwanda, Tanzania, and Uganda. It was originally established in 1967, with Kenya, Tanzania and Uganda as its only members, disbanded in 1977, and re-established in 2001.

A key purpose of this chapter is to provide the context and background of the secondary school education system in four of the partner states in the EAC; namely, Kenya, Rwanda, Tanzania, and Uganda.[1] The chapter begins with a brief historical account on formal education system in East Africa, with consideration of the economic situation. This is followed by a description of the structure of current school system in the post-independence East Africa, with some closing reflections on implications for harmonization.

Historical Background of Education in East Africa

The earliest system of education in East Africa was oral in nature. Its main objective was to ensure that the next generation inherited the culture, customs and rules of engagement in the society (Merriam 2007). In other words, its main objective was to mould individuals to fit into the society (Omari and Mosha 1987). A lot of emphasis was placed on communal virtues and collective responsibilities. Members of the larger kin network were collectively responsible for the upbringing of children and teaching to them values and norms of their family, society and culture. These values and norms were transmitted through various aspects of culture such as folk tales, songs, myths, legends, proverbs and riddles.

The philosophy of belonging, living together, and bravery was deep rooted in East African informal education. For example, among the Kiswahili speakers, there were proverbs that emphasized communalism; such as *mtu ni watu* (one can only be defined in the context of others; Mbiti 1970). The indigenous education also encouraged hard work using such proverbs as '*mtaka cha mvunguni sharti ainame*' (the one who desires that which is underneath must stoop; Moumouni 1968).

[1]While the issues covered in this volume are largely relevant to all the five partner states in the EAC, the in-depth focus does not cover Burundi.

The formal education system of the kind that is prevalent in East Africa today was established in the colonial period mainly through the Christian missionaries with the aim of directing "the conduct of individuals to respect and worship God" (Lawuo 1975). One of the key policy pronouncements that formalized the education system in East Africa was provided by the Phelps-Stokes Commission of 1919–1924. This commission emphasized that education provided to East African countries be adapted to the East African rural environment and required that all schools be registered, use the government prescribed syllabus, and be supervised by government inspectors (Masudi 1995). One of the shortcomings of the colonial education system was that it did not reflect the needs, interests and values of East Africans. Instead, its main aim was to prepare the 'natives' to be used in the administration of the colony, and to inculcate in them a liking for order, cleanliness, diligence, dutifulness, and a sound knowledge of the colonialists' customs and culture (Omari and Mosha 1987; Brock-Utne 2006). Most of the schools had three streams; namely (a) a stream for the 'bright pupils', who were mainly taught the 3Rs; i.e., reading, writing and arithmetic; (b) a stream for less academically able pupils, who were mainly taught industrial skills; and (c) a stream for the 'dull pupils', who were mainly taught agricultural labor skills (Gottneid 1976).

At independence (Kenya in 1963, Rwanda in 1962, Tanzania in 1961 and Uganda 1962), the four East African countries inherited an educational system that was mainly racially segregated: there were separate schools for those of European, Asian and African descent (Arunsi 1971; Galabawa 1990; Wasonga 2012). The education system for those of African descent was mostly in the hands of the missionary groups (Arunsi 1971), and had the stated aims of 're-socializing' them, facilitating evangelization, and meeting the colonial administrative objectives (Jimenez and Lockheed 1995; Omari and Mosha 1987). During the first years of independence, the East African countries were under enormous pressure to increase access to education.

A significant influence on education in post-independence East African states was the Addis Ababa Conference of Education for Africa in May 1960 (UNECA/UNESCO 1961). The conference observed the non-contextual nature of education provided by colonialists and came up with recommendations that the African education authorities should revise and reform education in the areas of curricula, textbooks and teaching methods, so as to take account of the African environment, child development, cultural heritage, and the demands of technological progress and economic development (UNECA/UNESCO 1961).

Cultural and Socio-Economic Background of the Countries

Before proceeding to consider the structure of education across the region more fully, the educational scene is put into the context of the cultural and socio-economic situation in which the countries operate.

Table 1.1 Selected socioeconomic indicators of East African countries. Norway and Niger have been included in the table to aid comparison

HDI[a] rank	Country	HDI value	Life expectancy at birth	Mean years of schooling[b]	Expected years of schooling[c]	Population (millions)	GDP[d] (billion USD)
1	Norway	0.944	81.5	12.6	17.6	5.0	512.58
147	Kenya	0.535	61.7	6.3	11.0	44.4	44.10
151	Rwanda	0.506	64.1	3.3	13.2	11.8	7.452
159	Tanzania	0.488	61.5	5.1	9.2	49.3	33.23
164	Uganda	0.484	59.2	5.4	10.8	37.6	21.48
180	Burundi	0.389	54.1	2.7	10.1	10.2	2.718
187	Niger	0.337	58.4	1.4	5.4	17.8	7.356

Source UNDP (2014), World Bank (2014)
[a]The Human Development Index (HDI) is a value assigned to a country between 0 and 1, giving a measure for average life expectancy, education and personal income. For more information see http://hdr.undp.org/en/content/human-development-index-hdi
[b]Mean years of schooling: years that a 25-year-old person or older has spent in schools
[c]Expected years of schooling: years that a 5-year-old child is expected in spend in education over a lifetime
[d]Gross Domestic Product

The East African Community (EAC) region covers an area of 1.8 million square kilometers with a population of about 143.5 million (East African Community 2014). The region reflects diversity of culture, language, ethnic and religious identities. As can be seen in Table 1.1, the four countries are not the poorest in Africa, but in terms of life expectancy, schooling and gross domestic product they fall well below Norway which, according to the Human Development Index, comes at the top of the list.

Educational Structure in the EAC

In terms of the structure of the education system, the East African countries continued with the structure they inherited from their colonial masters. For example, between 1967 and 1977, Kenya, Uganda and Tanzania had a single system of education under the East African Community—consisting of 7 years of primary education, 4 years of ordinary level secondary education (O-level), 2 years of advanced level secondary education (A-level) and at least 3 years of university education; that is, 7-4-2-3+. When the East African Community collapsed in 1977, each of the three countries continued with the same system of education. In 1985, however, Kenya adopted a new education system consisting of 8 years of primary education, 4 years of secondary education and at least 4 years of university education; that is, 8-4-4+. To date, Tanzania and Uganda are still using the post-1977 system of education. The only difference between the two systems of education is in the number of years of pre-primary education—two years

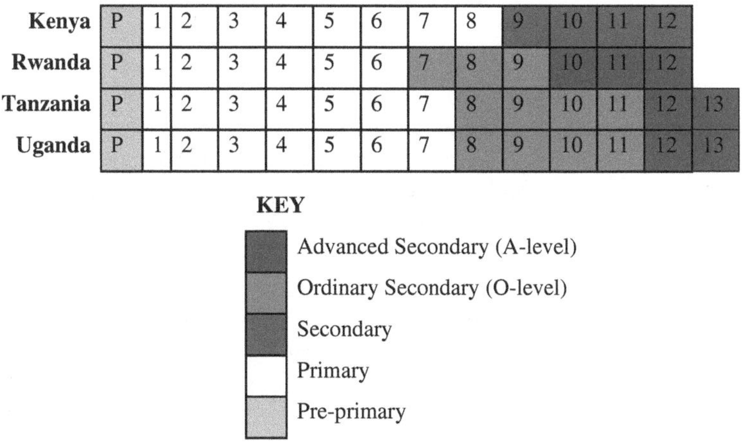

Fig. 1.1 The structure of pre-university education in Kenya, Rwanda, Tanzania, and Uganda

in Tanzania and three years in Uganda. As noted above, Rwanda is a member of the new East African Community. Her education system has also evolved since independence, and currently follows a 3-6-3-3-4+ structure; that is, 3 years of pre-primary education, 6 years of primary education, 3 years of ordinary secondary education (O-level), 3 years of advanced secondary education (A-level), and at least 4 years of university education.

Figure 1.1 shows the structure of pre-university education in the four countries. Excluding pre-primary education, Kenya and Rwanda have 12 years of pre-university education, while Tanzania and Uganda have 13 years. In Kenya and Tanzania, the official secondary school-entrance age is 14 years, while in Rwanda and Uganda it is 13 years (East African Community 2012).

In all the four countries, pre-primary (nursery) education is neither free nor compulsory. Pre-schools are either public-, private- or community-owned, and admission into these schools depends on the willingness and ability of the families to meet the school requirements. Primary education, which officially starts at the age of 6 years in Kenya and Uganda and 7 years in Tanzania and Rwanda, is free and compulsory in all the four countries. Whilst the first three years of secondary education are free and compulsory in Rwanda, in the other countries with the exception of Zanzibar[2] secondary education is not compulsory, with admission depending on passing the required test (East African Community 2012).

[2]Zanzibar is semi-autonomous with some aspects of education coming under the control of the Revolutionary Government. In a transition to compulsory secondary education for all, most students who fail the Primary School Leaving Examination are required to repeat the final year of primary school and retake the examination. Those who then fail leave school, whilst all who pass are required to proceed to the first two years of secondary school. After two years of secondary school there is another examination, those who pass are required to stay for two more years, those who fail then leave (see http://www.moez.go.tz/index.php?cq=syst&sys=3 retrieved on 26th February 2015).

Concluding Reflections

The formal educational system in East Africa has gone through several reforms over the years. In contrast to the colonial education system, the post-colonial education system in East Africa is aimed at equipping East Africans with the skills they need to compete nationally, regionally and internationally. Challenges, however, abound in this endeavor, and harmonization of the education systems of the partner states in EAC is seen as an important counterweight to these challenges. With this harmonization, it is envisaged that there will be easy students' mobility and easy employability throughout the EAC partner states. In recognition of the critical role that mathematics, science and technology play in the development of a country, more priority has been given to the harmonization of the curricula of these subjects (World Bank 2011). The work reported in this volume is expected to offer valuable insights on how best to go about the harmonization process.

References

Arunsi, N. O. (1971). The library and adult education in Tanzania: A survey. In A.-B. Wallenius (Ed.), *Libraries in East Africa* (pp. 83–124). Uppsala: Scandinavia Institute of African Studies.

Brock-Utne, B. (2006). *Whose education for all? The recolonization of the African mind.* San Diego, CA: Africana Books.

East African Community (2012). *Regional report: Harmonisation of the East African education systems and training curricula.* Retrieved January 24, 2015 from http://www.eac.int/education/index.php?option=com_content&id=73&Itemid=145.

East African Community (2014). *East African Community facts and figures—2014.* Retrieved January 23, 2015 from http://www.eac.int/statistics/index.php?option=com_docman&Itemid=153.

Galabawa, C. J. (1990). *Implementing educational policies in Tanzania.* Washington, D.C.: World Bank.

Gottneid, A. J. (Ed.). (1976). *Church and education in Tanzania.* Nairobi: East African Publishing.

Jimenez, E., & Lockheed, M. (1995). *Public and private secondary education in developing countries: A comparative study.* Washington, DC: The World Bank.

Lawuo, Z. (1975). *The implementation of British colonial education: Policy in Tanganyika, 1919–1939.* Dar es Salaam: University of Dar es Salaam Press.

Masudi, A. (1995). The role of donors and non-governmental organizations in primary education. *Papers in Education and Development, 16*, 105–116.

Mbiti, J. S. (1970). *African religions and philosophy.* London: Heinemann.

Merriam, E. O. (2007). *Learning in adulthood: A comprehensive guide.* San Francisco, CA: Jossey-Bass.

Moumouni, A. (1968). *Education in Africa.* New York: Praeger.

Omari, I. M., & Mosha, H. J. (1987). *The quality of primary education in Tanzania.* Nairobi: Man Graphics Limited.

UNDP (2014). *UN human development report, 2014.* Retrieved September 6, 2014 from http://hdr.undp.org/sites/default/files/hdr14-report-en-1.pdf.

UNECA/UNESCO (1961). *Final report: Conference of African states on the development of education in Africa.* Addis Ababa: UNESCO Publishing.

Wasonga, T. A. (2012). Perspectives on K-12 learning and teacher preparation in East Africa. In O. N. Ukpokodu & P. Ukpokodu (Eds.), *Contemporary voices from the margin: African educators on African and American education* (pp. 129–148). Charlotte, NC: Information Age Publishing.

World Bank. (2011). *A regional exploration of pathways towards harmonization of math and science curriculum in the East Africa Community: Discussion paper*. Washington, D.C.: World Bank.

World Bank (2014). *Countries' reports*. Retrieved September 6, 2014 from http://www.worldbank.org/en/country.

Chapter 2
Towards a Harmonized Curriculum in East Africa: A Comparative Perspective of the Intended Secondary School Mathematics Curriculum in Kenya, Rwanda, Tanzania and Uganda

Simon Karuku and Geoff Tennant

Abstract This chapter presents the results and insights gained from a comparative analysis of the national secondary school mathematics syllabuses of Kenya, Rwanda, Tanzania and Uganda. Whilst considerable commonalities were found, it was particularly Rwanda, with a different colonial past, which was found to have a more formal curriculum, exemplified in the lack of mention of measurement below the fourth year of primary school and the absence of approximation as a topic. Differences in sequencing topics were found, e.g. Uganda and Rwanda introduce number sequences in year 4 as a means of contextualizing algebra, Tanzania and Kenya introduce them in year 10 somewhat more formally. The continuance of calculations in shillings and cents in Tanzania and Kenya, and the use of logarithms as a calculation tool in Tanzania, Uganda and Kenya can be taken as evidence of a mismatch between what happens in school and the outside world. Differences are acknowledged between the stated curriculum on the one hand, and what gets taught by teachers and indeed learnt by children on the other. Whilst considerable work has been undertaken already to update the curricula in different ways, further work is required to ensure that the syllabus across the region is fit for purpose for the 21st century, not least in engaging all learners in this important subject.

S. Karuku (✉)
Embu University College, P.O. Box 6, Embu 60100, Kenya
e-mail: simon.karuku@embuni.ac.ke

G. Tennant
Aga Khan University Institute for Educational Development East Africa,
Salaama House Urambo Street, PO125, Dar es Salaam, Tanzania
e-mail: geoff.tennant@aku.edu

A. Halai and G. Tennant (eds.), *Mathematics Education in East Africa*,
https://doi.org/10.1007/978-3-319-27258-0_2

Introduction

Mathematics education in the East African region, just like in many parts of the world, faces many challenges and opportunities associated with population growth, technological advancement, regional interdependence, and globalization. In response to these challenges and opportunities, several curricular reforms in mathematics education have been proposed and implemented in the region since independence. The latest of these endeavours is the proposal to have a harmonized curriculum across the partner states in the East African Community (EAC). In a bid to contribute to the ongoing regional debate on how to harmonize the curriculum, a comparative analysis of the national secondary school mathematics syllabuses of four partner states in the EAC—namely, Kenya, Rwanda, Tanzania and Uganda—was carried out. In this chapter, the results of this comparative analysis, as well as the insights drawn from the analysis are discussed. At the end of the chapter is a discussion of the potential implications of the commonalities, differences and gaps identified in the syllabuses for practice, policy, and future research directions.

Before proceeding further, however, it needs to be acknowledged that there are, broadly speaking, three ways of considering the curriculum: what is intended by policy makers and the authors of the stated curriculum; what is taught by teachers in classrooms; and what actually gets learnt by students (Cuban 1992). One might reasonably consider that the most crucial aspect of the curriculum is the last, that is, what is actually learnt—Tennant and Sarungi in this volume use achievement scores as a proxy for what actually gets learnt by students. This chapter is concerned with one aspect of the intended curriculum; namely, the content of secondary school national mathematics syllabuses in the four countries. In the following section, we describe the intended aims of mathematics education in the four countries.

Stated Aims and Objectives of Secondary School Mathematics Education

In all the four countries, the mathematics curriculum is developed centrally by an independent governmental agency within the Ministry of Education. In Kenya, the curriculum is designed by the Kenya Institute of Curriculum Development; in Rwanda and Uganda, the agency responsible for curriculum development is known as National Curriculum Development Centre (NCDC), while in Tanzania, it is known as Tanzania Institute of Education (TIE). In each of the four countries the mathematics syllabus specifies what topics must be learnt. It also suggests the teaching, learning and assessment materials that teachers could use in their classrooms.

An analysis of the stated aims and objectives of secondary school mathematics curriculum found that some of the aims are country-specific, while others are more general. For example, one of the general objectives for A-level mathematics education in Rwanda is to enable the learner to "Use acquired mathematical skills to respect human rights" (NCDC–Rwanda 2010: 5)—a recognition of the potential role that mathematics can play in the realm of citizenship and community building. Some goals are shared by some or all the four countries. For example, there are two general objectives that are common to all the four countries, namely, to enable students to:

(i) think and reason precisely, logically and critically in any given situation, and;
(ii) creatively and confidently apply mathematical problem-solving strategies to solve problems in other subject areas and in everyday life.

In addition, there is one general objective that is shared by Kenya, Uganda and Rwanda, namely, to enable students to analyse, interpret and present numerical information meaningfully using tables, graphs, charts, and diagrams. These common objectives could serve as the initial reference points for the harmonization process.

The analysis also revealed that different terminologies have been used by different syllabuses to describe the purpose of teaching mathematics at the secondary school level. Whereas Kenyan and Rwandan syllabuses use only two terms (general objectives and specific objectives), Tanzanian and Ugandan syllabuses use five different terms. In particular, the Tanzanian syllabus uses general competences, class competences, class objectives, general objectives, and specific objectives, while the Ugandan syllabus uses aims, competences, general objectives, specific objectives, and learning outcomes.

In some instances, there seems to be lack of clarity on how the different terms are being used in the syllabuses. For example, in the Ugandan O-level Mathematics syllabus, specific objectives are used to describe what the learners are to achieve by the end of a topic, in the Principal Mathematics syllabus, general objectives and specific objectives are used for topics and subtopics, respectively. In the Subsidiary Mathematics syllabus, meanwhile, learning outcomes and competencies are used for topics and subtopics, respectively. In the Rwandan A-level syllabus, general objectives are used to describe what the learners are to achieve by the end of A-level, by the end of a given year, and by the end of a given topic, while specific objectives are used to describe what the learners are to achieve by the end of a subtopic. In the O-level syllabus, there are no general objectives for topics, instead specific objectives are used to cater for both topics and subtopics.

In moving towards harmonization, it would be helpful to have clarity in the nomenclature used, so that the syllabus makes it clear what is expected of teachers and students and why. This issue is reconsidered in the conclusions. Furthermore, there is need to ensure that the aims and objectives of the mathematics curriculum keep pace with the competencies required in the regional labour market. For instance, ICT skills are very much needed in the current job market, and it would thus be good to have some of the aims and objectives reflect this reality.

Organizing Principles Embedded in the Mathematics Syllabuses

We compared the stated or implied philosophy of secondary mathematics teaching and learning in the four syllabuses, and observed that, generally, the underpinning philosophies in Kenya, Tanzania and Uganda reflect an emphasis on real-world applications of school mathematics. The Kenyan syllabus "has emphasized application of Mathematics to real life experiences and practical approaches to teaching and learning in an effort to address such contemporary issues as information technology, health, gender and integrity" (KIE 2002: 3). The Tanzanian syllabus emphasizes a paradigm shift from a content-based to a competency-based curriculum. The syllabus states that the teaching and learning processes should be student-centered and activity oriented (TIE 2005b: iii). The Ugandan O-level syllabus encourages "practical problem solving" and cooperative learning (NCDC–Uganda 2008: x), while the A-level syllabus emphasizes "effective and efficient acquisition of knowledge and development of concepts, skills, values and attitudes" (p. 119).

Meanwhile, the Rwandan syllabus appears to be content-oriented and prescriptive. In the general introduction, the key topics of the syllabus are listed. The O-level syllabus approaches each topic "in a systematic way while taking into account difficulties it presents and the learning age of the student who receives it" (NCDC–Rwanda 2006: 4), while the A-level syllabus has been prepared "in a format which helps teachers to teach a particular topic effectively" (NCDC–Rwanda 2010: 4). The syllabus underscores the need for an interdisciplinary approach to the teaching of mathematics; in other words, the need to "establish the relationship between Mathematics and other subjects" (p. 7).

In practice, it may be helpful to think of content and process not as antagonistic but, rather, as two sides of the same coin. As such, the match between the stated underpinning philosophy and what happens in the classroom would appear to be a very fruitful area for future research to pursue.

The Place of Mathematics in the Overall School Curriculum

In all the four countries, mathematics has a central and significant place in the school curriculum, as reflected in the number of hours allocated to mathematics (relative to the other subjects in the school curriculum) and the requirement for mathematics to be compulsory for some of the years of schooling.

The secondary education curriculum in Kenya emphasizes both compulsory and elective subjects. In the first and second years, students study 12 subjects whereas in the third and fourth years, they specialize and study between 5 and 9 subjects. Mathematics, English and Kiswahili are compulsory at all levels of secondary education. Performance of students in mathematics and English influences their admission to various specialized programmes and courses in post-secondary institutions, thus determining their career path (Wasanga, Ogle, & Wambua 2012: 3). In Uganda (Karp, Opolot-Okurut, & Schubring 2014), Tanzania (TIE 2005a) and Rwanda (Republic of Rwanda 2013), mathematics is compulsory up to the end of ordinary secondary education. This means that, whilst Kenyan students have 12 years of compulsory mathematics education, their counterparts in Uganda and Tanzania have 11 years, while students in Rwanda have 9 years. In Tanzania, students who wish to study more mathematics at O-level can take "Additional Mathematics" in addition to the regular mathematics curriculum. The curriculum for additional mathematics is intended for the entire O-level cycle.

During the years of compulsory mathematics education, Kenya is the only country among the four under consideration that has two alternative routes for the secondary mathematics curriculum. Students joining secondary education can choose between 'Mathematics Alternative A' and 'Mathematics Alternative B' curriculum. The latter—a simpler version of Mathematics Alternative A curriculum—was developed in 2009 for students for whom secondary education is terminal or those whose career paths do not require intensive use of mathematics (KIE 2008). The mathematics curriculum is organized into topics and whereas Mathematics Alternative A curriculum has 68 topics, Mathematics Alternative B curriculum has 57 topics. The only new topic in Mathematics Alternative B curriculum is *Cash Books*, otherwise all the other topics are taken from Mathematics Alternative A curriculum. However, as noted in the next chapter, the uptake of Alternative B is extremely low, and it would appear in practice that there is still work to be done in ensuring a meaningful, challenging and enjoyable mathematics curriculum for all learners.

In Tanzania and Uganda, A-level students select a combination of three principal subjects to specialize in according to their interests and career goals. However, students who do not take mathematics as one of their principal subjects but have at least one principal subject that requires some mathematical background—for example, economics or physics—are required to take subsidiary mathematics. This subsidiary mathematics is not as rigorous as the one offered as a principal subject. Until 2014 in Tanzania 'Basic Applied Mathematics' was assessed just on a pass-fail basis, this continues to be the case with the equivalent in Uganda.

It is worth noting here that in Uganda, a project to reform the O-level curriculum is currently underway. According to the National Curriculum Development Centre (NCDC), the proposed curriculum seeks to shift from a strictly academic list of subjects to a set of generic skills that are to be acquired through eight compulsory learning areas; namely, Mathematics, Science, Languages, Social Studies, Creative Arts, Life Education, Religious Education, and Technology and Enterprise (NCDC-Uganda 2014). Currently, the O-level curriculum consists of 14 subjects, and the proposed curriculum framework seeks to merge and integrate these subjects. It is hoped that the new curriculum will provide learners with holistic education that will equip them with the requirements for the contemporary market. With respect to the mathematics learning area, the new curriculum will "move to the applicable and functional mathematics that is required by all learners for full and effective participation in social and economic life" (p. 3). There will be a core programme of study for all learners and an extended programme designed specifically for high achievers and those going on to further studies in mathematics. A comparison of the current curriculum (NCDC-Uganda 2008) and the proposed curriculum (NCDC-Uganda 2014) would indicate that a number of topics are being dropped, including logarithms as a calculation aid, sets, matrices, functions and linear programming. Algebra is addressed with a noticeably more problem-solving approach. In the absence of any clearly stated timelines for the implementation of the new curriculum, it is the current curriculum which is subject to analysis in this chapter.

Analysis of the Subject Matter Content

Our analysis of the subject matter content in the syllabuses focused on five content strands in mathematics, namely, Number Sense and Operations, Algebra, Geometry, Probability and Statistics (Data Handling), and Measurement. In our analysis of each of these content strands, we were interested in the breadth (the content specified in each of the key ideas) and depth (the degree of emphasis given to the key ideas and the skills specified in relation to those ideas). We were also interested in the grade level at which specific content is introduced. Because of the continuum in mathematics from primary to secondary school, it was necessary to analyze the subject content in the pre-secondary school education as well. Based on this analysis, we identified the disparities and similarities between the syllabuses of the four countries. This section presents the results of the analysis of each of these four content strands.

It is worth noting that it is not always clear when one topic ends and another begins. For example, number sequences might be considered as either a Number Sense and Operations or an Algebra topic; similarly, quadratic equations can be considered both under the headings of Algebra and Geometry. In addition, there are instances where topics are explicitly mentioned in one country but not another, leaving a gap in the intended curriculum which may well be filled in practice. So, for example, place value is not explicitly mentioned in the Rwanda syllabus, and algebraic substitution is not explicitly mentioned in the Tanzanian syllabus. It is also important to reiterate that this was an analysis of the intended curriculum, not the taught curriculum, and still less the learnt curriculum.

Number Sense and Operations

There is a great level of commonality among the four syllabuses in terms of the breadth of coverage of the key concepts in Number Sense and Operations. However, there are differences in depth of coverage as well as in the grade level at which some of these key concepts are introduced. Table 2.1 shows the year when some of the key skills in Number Sense and Operations are introduced in the mathematics syllabus.

It is noticeable particularly that different syllabuses have different emphases on approximation and estimation. For example, the Kenyan syllabus considers significant figures, absolute, relative, percentage, round-off, and truncation errors, propagation of errors from simple calculation, and maximum and minimum errors. The Ugandan syllabus, on the other hand, focuses on significant figures, decimal places, 'rough' answers, and sensible answers. The Tanzanian syllabus focuses on rounding off whole numbers to given place values; rounding off decimal numbers to a given number of decimal places, writing numbers to a given number of significant figures, and computations involving large and small numbers. For example, in Year 7, there is a discussion of the rounding-off of whole numbers to given place values or to a given number of significant figures. The Rwandan syllabus, meanwhile, has no explicit mention of this area of Number Sense.

It is noticeable that logarithms are still taught in Kenya, Tanzania and Uganda as a calculation aid, whereas in Rwanda they are introduced in the context of calculus. Furthermore, it is only in the Tanzanian mathematics curriculum that the use of ICT and calculators in mathematics is included in the curriculum; otherwise the other three countries do not make explicit mention of integration of ICT in mathematics.

Table 2.1 Year when various skills in number sense and operations are introduced

	Year when the skill is introduced			
	Kenya	Rwanda	Tanzania	Uganda
Approximation and estimation	11	Not covered	8	8
Number line	9	Not covered	6	1
Fractions	3	2	1	1
Integers	9	7	6	5
Place value	1	Not explicitly covered. There's reference to ones, tens, hundreds, etc.	8	1
Number bases	Not covered	5	8 (only base 10)	5
Roman numbers	5	6	4	4
BODMAS	Not explicitly covered		7	6
Order of operations	9 (on fractions, integers, decimals)		8	
Commutative, associative and distributive laws	Not covered	3	Not explicitly covered	Only commutativity in year 3
Logarithms	10	10	9	9

Algebra

The following shows the introduction of algebra topics by Year across the region, without those topics in which the introduction was in the same Year in all countries (Table 2.2).

Table 2.2 Year when various skills in algebra are introduced

	Year when the skill is introduced			
	Kenya	Rwanda	Tanzania	Uganda
Use of letters to represent numbers	4	4	5	5
Simplify algebraic expressions	5	6	5	5
Use of > and < symbols	6	8	8	7
Use of brackets	6		9	Not explicitly mentioned
Find value of algebraic expressions through substitution	7	6	Not explicitly mentioned	6
Simplify inequalities in 1 unknown	7	8	8	9
Form and simplify algebraic expressions	8	7	8	9
Factorize algebraic expressions	9	8	9	9
Remove brackets	9	8	9	
Apply algebra to real life situations	9	Not explicitly mentioned	8	3
Solve simultaneous equations	9	9	8	10
Form and solve linear equations in 1 unknown	6	6	6	6
Form and solve linear equations in 2 unknowns	9	(10)	8	10
Form and solve quadratic equations	10	(10)	9	9
Form and solve linear inequalities	10	(10)	Not explicitly covered	11
Transform formulae, constants of proportionality, use to solve problems	10	Not covered	Not covered	10
Arithmetic and geometric progressions	10	4	10	4
Binomial expansions	10	Not covered	Not covered	Not covered
Linear programming	11	Not covered	11	11
Use of matrices to solve simultaneous equations	11	Not covered	11	Not covered
Forming sets (including the empty set)			1	

(continued)

Table 2.2 (continued)

	Year when the skill is introduced			
	Kenya	Rwanda	Tanzania	Uganda
Comparing sets			2	
Sets: Venn diagrams, membership, inclusion	9	5	5	
Union of sets	9	6	5	
Sets: application to probability			6	
Finite and infinite sets			7	
Sets: formal notation	9	7	8	
Functions: injection, subjection, bisection (sic)		7		
Functions: domain and image	10	7		
Functions: partition		7		
Functions: inverses	10		11	
Functions: composite functions including formal notation		7	11	
Number systems with formal notation: integers, decimals, rationals		7		
Number systems: real		8		
Groups rings and fields		8		
Introductory differentiation	12	(11)	11 (but informally)	
Integration: definite integration	12	(12)	11	
Differential equations	12	(12)		

Note Years in brackets are those not compulsory in Rwanda

There are a large number of commonalities in the sequencing of algebra topics across the countries. For example, letters to represent numerical values is first introduced in Year 4 in Kenya and Rwanda and Year 5 in Uganda and Tanzania. There is similarly at most one year's difference as to when simplification of algebraic expressions are introduced (5 or 6), solution of equations in one unknown (6), finding values of algebraic expressions through substitution (6 or 7), solving quadratic equations by factorization and completing the square (9 or 10), forming and solving linear inequalities (10 or 11) and linear programming, with two years' discrepancy in solving simultaneous equations (Tanzania in Year 8, Rwanda and Kenya in Year 9, Uganda in Year 10).

One key difference is the positioning of number sequences. In Uganda and Rwanda this is in year 4, in Kenya and Uganda in year 10. Particularly in Uganda, there is the clear implication that number sequences are intended to be a means of starting to develop algebraic ideas then to formalize these ideas later. This is not so clear in Rwanda. However, in Kenya and Uganda number sequences are treated very formally from first introduction, consistent with the positioning at a later stage in the school curriculum.

Further differences arise with sets. Forming sets, including the empty set, is mentioned in the Ugandan syllabus in Year 1, with comparison of sets in Year 2. Meanwhile, explicit mention of sets is first made in Uganda and Rwanda in Year 5 and in Kenya in Year 9. In Rwanda, a high level of formality is noticeable from a very early stage. So, for example, 'bijection' comes up in year 7, with groups, rings and fields coming up in year 8. Whilst A-level in Rwanda starts at Year 10, with not all pupils staying on at that stage and not all those that do stay on doing mathematics, the sense of complexity in Rwandan topics continues with internal composition law for groups, rings and fields referred to in Year 10, and the concept of continuity within analysis in Year 11.

Geometry

Table 2.3 shows the year of introduction of geometry topics across the region.

As is the case with the Algebra strand, there is a large amount of commonality, particularly in higher grades, in the treatment of the Geometry strand across the four countries. Angles are first mentioned across the countries in Year 4, with problems with properties of quadrilaterals coming up in Year 7 or 8. In general, there is at most a 1 year difference in application of trigonometry to lengths of sides of right angled triangles (Years 9 and 10) and areas of triangles (Years 10 and 11). Similarity and enlargement are addressed in Years 9 and 10, as are reflection and congruence and areas of parts of circles.

At the early stages of the curriculum, there are large apparent differences as to when topics are introduced. For instance, rectangles, triangles and circles are explicitly mentioned in the Kenyan and Ugandan syllabuses in Year 1, in Tanzania in Year 2 but not until Year 4 in Rwanda. Consistent with this difference is the suggestion in Kenya, Tanzania and Uganda that these shapes be used to make patterns in Years 2 and 3, there is no such suggestion in the Rwandan syllabus.

As with the Algebra strand, a greater formality can be seen in Rwanda, with scalar products in the treatment of vectors in Year 10 and cross product in Year 11. Similarly, whilst calculations in 3 dimensions are explicitly mentioned in the Rwandan syllabus in Year 7, these come rather later in the other countries.

Probability and Statistics

There are noticeable differences among the four countries in the breadth of coverage of the Probability and Statistics strand. In general, the Ugandan and Rwandan syllabuses, on the one hand, cover more key concepts than the Kenyan and Tanzanian syllabuses, on the other.

Differences are also evident in terms of the grade level at which key concepts in this strand are introduced. For example, the Ugandan thematic curriculum includes

Table 2.3 Year when various skills in geometry are introduced

	Year when the skill is introduced			
	Kenya	Rwanda	Tanzania	Uganda
Straight and curved lines	1		4	
Estimate capacity using containers				1
Rectangles, triangles, circles	1	4	2	1
Ovals	2			
Draw map			3	
Area and perimeter of shapes			3	
Measure angles	5	5	6	5
Reflection (line symmetry)			5	5
Rotations				5
Equilateral, isosceles and scalene triangles	5	4	4	
Parallel lines	5	4	5	6
Properties of rectangles and squares	5			6
Construction for drawing perpendicular and parallel lines, bisecting a line	6		8	7
Vertically opposite and supplementary angles	6	5	8	7
Construction for bisecting angles with applications	6	5	8	7
Construction for triangles	6	6	8	7
Circles: constructions and naming of parts	6		6	
3D: identify cube, cuboid, sphere, cylinder		5		3
3D: surface area and volume		5		
3D: edges, faces, vertices of cubes and cuboids	6		6	9
Circles: pi and circumference and area		6	8	
Construction circle from vertices of triangle	7	7		
Pythagoras	7	Not explicitly mentioned	7	6
Models of cubes, cuboids, cylinders	7			
Construct circle touching edges of triangle	8			
Problems with Pythagoras	8			
Construct parallelograms and rhombuses	8			
Properties of quadrilaterals	8	7	8	
Triangular and square based pyramids	8			
Cartesian plane, including graphical solution of simultaneous equations	9	8	6	8
Parallel projections, central symmetry		8		
Solve problems with angles	9			8
Constructions including bisector of lines, parallel lines, regular and irregular polygons	9			
Reflection (line and plane symmetry)	10			

(continued)

Table 2.3 (continued)

	Year when the skill is introduced			
	Kenya	Rwanda	Tanzania	Uganda
Rotation	10	8	9	
Translation			9	
Surface areas of solids	10			
Volumes of solids	10		6	
Trigonometry: all angles	11			9
Circles: lengths of arcs and chords; tangents; traversals	11	8 (Thales theorem)	10	8
Vectors	11	8	11	9
Solve quadratic and cubic equation graphically	11			8
Loci	12			8
Graphs of form sin(ax + b)	12	10		11
Calculations in 3D	12	7	11	9
Great and small circles	12		10	
Longitude and latitude	12		10	
Midpoints on lines			11	

basic bar charts in Year 2, while in the other countries, these are covered much later. Similarly, probability is introduced in year 7 in Uganda, 9 in Kenya, 11 in Tanzania and 12 in Rwanda.

There are also differences in the terminologies used to refer to this strand in the Ugandan syllabus at different stages of the mathematics curriculum. For example, during the primary school years, it refers to it as "Data Handling" or "Interpretation of Graphs and Data"; while at the secondary school level, it is called "Statistics". Table 2.4 shows the results of our comparative analysis

Table 2.4 Year when various skills in data handling/statistics are introduced

	Year when the skill is introduced			
	Kenya	Rwanda	Tanzania	Uganda
Collect and organize/record data	4	6	9	2
Bar graph/bar chart	5	6	4	2
Pie chart	10	6	6	7
Line graph	6	Not mentioned	6	4
Mean/average	6	5	5	5
Median	8	8	10	6
Mode	7	5	10	6
Quartile	12	9	(13)	(12
Range	12	(10)	Not covered	(12)
Standard deviation	12	(10)	(13)	(12)
Interquartile range	12	(10)	Not covered	(12)
Probability	9	(12)	11	7

of the syllabuses in terms of year when some of the key skills in Statistics and Probability are introduced in the mathematics curriculum.

In Tanzania, quartiles and percentiles are presented in year 13 as measures of dispersion (TIE 2009a: 48) when, in fact, they are not; they are measures of position/location. In the Subsidiary Mathematics syllabus, quartiles and percentiles are presented as measures of central tendency (TIE 2009b: 22).

Measurement

Compiling the table below necessitated, rather more than in the other content areas, making judgements as to the equivalency of wording in the different syllabuses. Gaps in the table indicate a lack of explicit mention, which does not, of course, necessarily mean that the area is not covered.

As shown in Table 2.5, there are very few similarities in measurement between all four countries.

Distance/time graphs get introduced in either Years 6 or 7 and time in hours and minutes appear in Year 4. Apart from these examples, it is the differences between the countries which are most noticeable.

There is no mention of measurement topics in the Rwandan syllabus before Year 4, with no suggestion that, for example, informal units be used to measure length as there is in the other three countries.

Consistent with the thematic curriculum in Uganda in the early stages of the primary school, measurement topics are well represented in the first three years, with Kenya also showing a large number of informal topics.

Money calculations are not explicitly mentioned in the Rwandan syllabus, with differences in the timing of introduction in the other three countries, with adding and subtracting in units of 1000 shillings appearing in Year 3 in Kenya, 2 in Uganda and 1 Tanzania (albeit that 500 shillings are also mentioned in Kenya and Uganda at this time). It is, of course the case that the same numerical values have different monetary values in the different currencies, although there is no clear pattern to explain this.

It is noticeable also that somewhat different contexts are given at different stages for financial topics. So, invoices are mentioned in Year 4 in Tanzania but not followed up thereafter, double entry accounts in Tanzania in Year 10, whilst taxation as a topic appears in Kenya in Year 11 and Uganda in Year 10.

Implications of the Findings

A number of key points come out of the analysis above. It is particularly the Rwandan syllabus which, as noted above, is defined largely by content, which adopts a formal approach, excluding approximation of number and including

Table 2.5 Year when various skills in measurement are introduced

	Year when the skill is introduced			
	Kenya	Rwanda	Tanzania	Uganda
Comparison of people, weights and lengths				1
Non-standard units	1		3	1
Times of day	1			1
Capacity of containers	1			1
Months of the year			*1 (in language)*	1
Use coins	1	4		
Add/subtract shillings in 500 and 1000 denominations	3		1 (100 s TSh)	2
Use a calendar	2	4	5	2
Measure in length	4	4	4	4
Measure in time	3	4		2
Measure in litres	3	4		2
Make a simple budget				3
Personal timetable				3
Weight in g and kg			3	3
Generate an invoice	5 (a bill)		4	
Problems in time, distance, speed				5
Net mass, gross mass, tare				5
General calculations with money				5
Profit and loss				5
Convert from one currency to another	9		10	6
Temperatures	7			
Area in m^2 and cm^2	5	4		
12/24 h clock time			6	7
Timetables	9			7
Calculate volume	5	8	7	7
Commercial arithmetic				8
Density	9			
Bearings				8
Distance-time graphs	10			8
Simple and compound interest	6		8	8
Hire purchases and mortgages	7			10
Taxation	11			10
Double entry accounts, balance sheets			10	

topics in year 8 which would in other countries be considered to be A-level topics or even undergraduate. According to Uworwabayeho, Rubagiza, & Iyamuremye (2007) the previous Rwandan syllabus was overloaded, content driven, trying to engage all children in the secondary sector with what only the very highest attainers were doing in the USA. This analysis would seem to suggest that there is the need to revisit these issues in ensuring a syllabus which meets the needs of all youngsters in school for the 21st century, although such a suggestion needs to be cross-referenced against research as to pupils' experience within mathematics in Rwanda, both qualitative and quantitative.

The extent to which there are connections between different aspects of mathematics, particularly number and algebra, and also algebra and geometry, also varies from country to country. Uganda's thematic curriculum for the primary sector, with the spirit of it largely being continued in the proposed new secondary curriculum (NCDC-Uganda 2014) would, in principle appear to be a starting point for exploring connections between different aspects of mathematics and, indeed, with other subjects and 'real life'.

It would be consistent with a thematic or competency-based approach to question what actually matters in terms of content coverage. So, for example, one might reasonably consider that many geometry topics—for example, shape work, transformations, symmetry, loci, constructions, angles—can be introduced informally at a very early stage in the mathematics curriculum, with links made with art, shapes that one sees around the school room and beyond. A corollary of this is that one may not wish to put much significance on the difference in Year, for example, in which rectangles, triangles and circles are specifically mentioned, as a later introduction may well mean a more formal treatment, and it is possible that these concepts are mentioned at an earlier stage informally even if not on the curriculum.

From this perspective the differences in what is addressed at the early stage of the school curriculum could be considered to be arbitrary, that it is how topics are approached is more important than exactly what is approached and when. This would suggest the need for additional research as to children's experience in these topics, and how teachers can best be supported in ensuring that children have a clear sense of progression in an interesting and motivating mathematical environment.

In making strong links between the mathematics classroom and the outside world, the handling of finance could usefully be rethought across the region. There is clear opportunity to practise basic arithmetic routines through money calculations, and also to make links with other subjects and life outside of school. The need for children to be doing calculations in both shillings and cents in Kenya and also Tanzania could usefully be rethought in terms of ensuring that the mathematics taught in classrooms is aligned with that actually used in the outside world. Also, whether the examples of applications of financial topics given relate meaningfully to pupils' experience needs to be questioned in working towards a syllabus fit for all children in the 21st century.

Conclusions

As earlier noted, this was an analysis of the intended curriculum and as such, it needs to be supplemented with research into the taught and learnt curricula. Some of the issues emerging from this analysis call for a thorough debate among mathematics educators and researchers. For example, there is need to examine the position of logarithms as a calculation aid, particularly against a backdrop of the pressures of an overloaded curriculum and the rhetoric of the need to embrace ICT, including calculators, as a valuable tool for mathematics learning. There is also a challenge raised of how to ensure that the mathematics curriculum represents the needs of all learners and not just the few. This would lead to the question of how long mathematics should be made compulsory with, as noted earlier, different answers given across the region.

It is clear from this analysis that a great deal of work has been done in the four countries considered to enhance mathematics education and students' experiences in the classrooms. It may well be considered that a major benefit of harmonization across the region would be to harness these efforts to make mathematics relevant and meaningful, recognising the need to accommodate all learners and not just a few, in order that students' learning experiences be enhanced to the benefit of themselves, their communities, their countries and the region of East Africa.

References

Cuban, L. (1992). Curriculum stability and change. In P. W. Jackson (Ed.), *Handbook of research on curriculum* (pp. 216–247). New York: MacMillan.

Karp, A., Opolot-Okurut, C., & Schubring, G. (2014). Mathematics education in Africa. In A. Karp & G. Schubring (Eds.), *Handbook on the history of mathematics education* (pp. 391–403). Dordrecht, The Netherlands: Springer.

Kenya Institute of Education. (2002). *Secondary education syllabus* (Vol. 2). Nairobi: Ministry of Education, Science and Technology.

Kenya Institute of Education. (2008). *Secondary education curriculum: Mathematics alternative 'B' syllabus*. Nairobi: Ministry of Education.

National Curriculum Development Centre, Rwanda (2006). *Mathematics curriculum for ordinary level*. Kigali: Ministry of Education.

National Curriculum Development Centre, Rwanda (2010). *Advanced level mathematics curriculum for science combinations*. Kigali: Ministry of Education.

National Curriculum Development Centre, Uganda (2008). *Mathematics teaching syllabus: Uganda Certificate of Education (Senior 1–4)*. Kampala: Ministry of Education and Sports.

National Curriculum Development Centre, Uganda (2014). *Reforming the lower secondary curriculum*. Retrieved November 30, 2014 from http://fenu.or.ug/wp-content/uploads/2013/02/New-learninfg-areas-lower-secondary-curriculum.pdf.

Republic of Rwanda. (2013). *2012 education statistics yearbook*. Kigali: Ministry of Education.

Tanzania Institute of Education. (2005a). *Curriculum for ordinary level secondary education in Tanzania*. Dar es Salaam: Ministry of Education and Vocational Training.

Tanzania Institute of Education. (2005b). *Basic mathematics syllabus for secondary schools Form I–IV*. Dar es Salaam: Ministry of Education and Culture.

Tanzania Institute of Education. (2009a). *Advanced mathematics syllabus for advanced level secondary education: Form V–VI*. Dar es Salaam: Ministry of Education and Vocational Training.

Tanzania Institute of Education. (2009b). *Basic applied mathematics syllabus for advanced level secondary education: Form V–VI*. Dar es Salaam: Ministry of Education and Vocational Training.

Uworwabayeho, A., Rubagiza, J., & Iyamuremye, D. (2007). A review of mathematics and science education in Rwanda. In R. Barwell, K. Bishop, S. Erduran, A. Halai, D. Iyamuremye, T. Nyabanyaba, N. F. Rizvi, S. Rodrigues, J. Rubagiza, & A. Uworwabayeho (Eds.), *Implementing curriculum change: Literature reviews—South Africa, Rwanda and Pakistan* (pp. 59–70). Bristol, UK: EdQual RPC.

Wasanga, P. M., Ogle, M. A., & Wambua, R. M. (2012). *The SACMEQ III project in Kenya: A study of the conditions of schooling and the quality of education*. Nairobi: Kenya National Examinations Council.

Chapter 3
Achievement in Mathematics: Comparative Analysis from East Africa

Geoff Tennant and Veronica Sarungi

Abstract Against a perception that results in mathematics are low, this chapter examines available data on achievement across the region, including two datasets based on surveys in the primary sector—Uwezo and SACMEQ—and O levels and equivalent. Common themes which emerged were wide variations across countries and within countries, with youngsters from urban areas performing more highly than rural. Whilst most surveys showed girls performing less well than girls, this was not uniformly the case. Unclear trends need to be put into the context of increased enrollment in the secondary sector. Where examiners' reports were available, common issues which arose were lack of use of the methods required and engaging with multi-step problems expressed in words, consistent with problems in using English as a medium. No indication in the reports reviewed was given of problems with the questions themselves, despite a number of mistakes being uncovered. In working towards harmonization, a review is needed as to how assessment systems can most effectively support high quality learning in the region.

Introduction

In forming a clear sense of the state of mathematics education across the region, one key issue is achievement data. Treating it as a proxy for the extent to which the subject is learnt, analysis as to how youngsters achieve in examinations of different forms throws light, in principle, on the teaching and learning of the subject, and other related issues such as teacher training and government policy.

G. Tennant (✉) · V. Sarungi
Aga Khan University Institute for Educational Development East Africa, Salama House, Urambo Street, P O Box 125, Dar es Salaam, Tanzania
e-mail: geoff.tennant@aku.edu

V. Sarungi
e-mail: veronica.sarungi@aku.edu

A. Halai and G. Tennant (eds.), *Mathematics Education in East Africa*,
https://doi.org/10.1007/978-3-319-27258-0_3

This chapter examines assessment data in Tanzania (considering Tanzania Mainland and Zanzibar separately where necessary), Uganda, Kenya and Rwanda. Reflecting both what is available and also the continuum from primary to secondary mathematics, first regional datasets are considered, before then looking at secondary national examination information by country. Some overall issues in quality of mathematics education and in the efforts towards harmonizing education provision across the region are then considered with recommendations both in terms of strengthening the analyses that can be conducted from assessment data, and also looking to improve the achievement that the data represents.

Regional Assessments: Uwezo and SACMEQ

In recent years there have been two major regional assessments testing mathematical skills of learners across most of the East African countries. Both assessments, Uwezo and the tests administered by the Southern African Consortium for Monitoring Educational Quality (SACMEQ) involved primary school children although the former focused on the age of children (6–16 years) and so also included secondary learners. The results of these tests are included in this chapter for two reasons. First, they provide vital information about the mathematical competences that learners achieve in primary school and with which they transfer to secondary school. With a push towards universal secondary education (USE) in all East African countries, the state of students' mathematical skills as they complete primary school will strongly influence mathematics teaching and learning as well as achievements at secondary school. The second reason for including a description of these assessments is to provide models of how harmonization could be achieved across several countries with different education structures and curricula.

Uwezo

Uwezo is an East African initiative that aims to support improvement in literacy and numeracy. Uwezo has conducted household surveys to measure the performance in basic literacy and numeracy of children aged between 6 and 16 years of age on an individual basis in Kenya, Tanzania and Uganda since 2009. There have been three rounds of tests to date, disseminated in 2011, 2012 and 2013. Approximately a third of a million children were involved in each of rounds 2 and 3.

The Uwezo tests in each country assess the arithmetical competences required at the end of standard two for each country. As a result, there are slight variations in the numeracy tests, which reflect differences in the national curricula and variations in emphasis. Particularly, it might be considered that standard three students and older would be able successfully to sit tests on standard two material but, as the Table 3.1 shows, this is not in fact the case for the second round (third round data is still incomplete).

Table 3.1 Percentage of students in standards three and seven passing numeracy test for standard two in Kenya, Tanzania and Uganda

	Std 3	Std 7
Kenya	31.1	91.0[a]
Tanzania	36.9	84.4
Uganda	12.9	73.9

Uwezo East Africa (2012)
[a]In Kenya the students were in standard 8

Table 3.2 Percentage of children between 10 and 16 years passing Uwezo tests in the second survey according to gender and country

	Kenya	Tanzania	Uganda
Female	72.9	61.3	57.2
Male	70.0	59.9	56.6
Difference (female–male)	2.9	1.4	0.6

Uwezo East Africa (2012)

The Uwezo reports focused on a variety of aspects including gender. As can be seen in the Table 3.2, contrary to other data reviewed in this chapter, female students are achieving more highly.

There are wide variations within countries consistently in all surveys. For example, in 2012, the percentage of children between 9 and 13 who passed the numeracy test was largest in Dar es Salaam region with 74.7 % and lowest in Kigoma region with 41.6 %. Similarly, in Kenya in 2012 the percentage of children who could do class 2 subtraction in the provinces of Nairobi and North Eastern was 72.0 and 51.1 % respectively. It needs to be noted, however, that the nature of sampling—600 households per district with 30 enumeration areas per district with 20 households per area—means that confounding issues such as school status and age of children are not clearly accounted for. Nevertheless, the wide variations within a specific country are consistent with SACMEQ and the Ugandan National Assessment of Progress in Education considered below.

SACMEQ

Please note: in the following section, Zanzibar is treated as a separate entity from Mainland Tanzania, corresponding to the fact that educational policy up to secondary Form 2 (i.e. the 9th year of compulsory education) in Zanzibar is administered by the Revolutionary Government.

SACMEQ is a consortium which has administered a series of surveys on educational achievement in the primary sector in Southern African countries. The most recently published survey, SACMEQ III, ran from 2006 to 2011 in 15 countries, involving approximately 61,000 learners, 8000 teachers and 2800 school principals.

Data has been collected under a variety of headings from students in standard 6—i.e. the last year of primary school in most countries—and their teachers. In mathematics, test items were devised in accordance with the curricula of the participating countries, with categories ranging from 'pre-numeracy' to 'abstract problem solving'.

In considering findings from SACMEQ III, the average for SACMEQ II for the whole sample was set at 500 with a standard deviation of 100 with efforts made to ensure consistency of standards from one survey to the next. In analysing preliminary data from SACMEQ III (Hungi, Makuwa, Ross, Saito, Dolata, van Cappelle, & Vellien 2010) the following key points emerge, concentrating particularly on Kenya, Tanzania and Uganda but considering other countries for comparison where appropriate.

Big differences can be seen between countries, so, against a SACMEQ III average of 510, Uganda is the lowest at 481, Zanzibar at 490, Mainland Tanzania at 553 and Kenya at 557. In addition, wide variation can be seen between regions within countries. This is particularly noticeable in Kenya, so, at the extremes, Western has an average of 516, against Nairobi of 610. To a lesser extent this is true also in Uganda, with a difference of 50 points between the highest and lowest achieving region. Similarly big differences can be seen between rural and urban scores. So, against an average difference across SACMEQ of 40, Zanzibar is lower at 23, the others are approximately the same as the average.

The small, 4 point difference between the average for boys and girls (boys higher) across the whole of SACMEQ III is similar to the difference in Zanzibar (5 points), with Uganda at 9 points difference, whilst Kenya had a difference of 21 and Mainland Tanzania of 31. In each of these cases boys achieved on average higher than girls, although this is not true across all SACMEQ III, in Botswana, Mauritius, Seychelles and South Africa girls achieved more highly, in the Seychelles by 30 points.

As well as giving individual scores, SACMEQ III also considered the percentage of students reaching different levels of competency. Particularly, 'Basic Numeracy' is at Level 3 of 8, with students at this level able only to deal with one arithmetic operation at a time. The following percentages of students achieved at Levels 1–3 (i.e. at basic numeracy or below): 38.3 % in Kenya, 43.1 % in Mainland Tanzania, 74.9 % in Uganda and 63.4 % in Zanzibar.

National Assessments

Kenya

The Kenyan Certificate of Secondary Education (KCSE) examinations are overseen by the Kenyan National Examination Council (KNEC) and administered at the end of the four-year secondary school period. KNEC disseminates KCSE examination reports annually. These reports contain information about the number of candidates, overall mean score and proportion of males and females. Furthermore, there is a report for each subject and within that each paper if there is more than one examination.

Table 3.3 Mean score of candidates who sat for mathematics KCSE alternative A

Subject	2010	2011	2012
Mean score	23.06	24.79	28.66

Kenya National Examination Council (2012, 2013)

Table 3.4 Enrolment and mean scores by gender from 2010 to 2012

	2010		2011		2012	
	Female	Male	Female	Male	Female	Male
Enrolment	157,816	196,055	181,770	228,117	195,093	241,233
Mean score	19.71	25.75	21.00	27.80	25.30	31.38

Kenya National Examination Council (2012, 2013)

Mathematics is the subject taken by most candidates. There are two papers each carrying 100 % and the final mark is obtained by taking the average. In 2010, KNEC introduced two alternatives for Mathematics, Alternative A taken by the vast majority of candidates and Alternative B taken by a very much smaller number of candidates achieving at considerably lower levels. The analysis below, therefore, considers only Alternative A.

Table 3.3 shows the mean scores for Mathematics Alternative A.

It is noteworthy that alongside the low mean scores the standard deviation is high. For example in 2011, the standard deviation for Alternative A was 22.15. This means that there is considerable variation in the scores achieved, consistent with a large number of students achieving in single figures.

Table 3.4 shows the enrolment and mean scores by gender from 2010 to 2012.

As can be seen, enrolment is increasing with consistently more male than female students attempting the examination—although it is worth noting that the number of male students in 2010 is almost exactly the same as the number of female students in 2012, so differences in uptake may not be entrenched. Even though there are more male students attempting the examinations, their mean score is higher, indicating that there is work still to be done to ensure gender equity.

Analysis Based on Examiners' Reports

Examination reports by KNEC usually indicate questions that were not done well or were unpopular by candidates. Students' problems can be generally classified into five categories that are not mutually exclusive.

- demonstrating the specific method for a given sub-topic, e.g. the trapezium rule
- applying mathematical concepts to real-life situations, e.g. most candidates were unable to realize that a word problem question required a geometric progression for its solution
- confusing terms that are closely related, e.g. simple interest and compound interest, and also decimal places and significant figures

- forming equations from word problems, e.g. in the context of linear programming. However, no problem was reported in candidates forming equations from word problems involving commercial arithmetic
- failing to follow the required steps prescribed by the question, e.g. with a calculator available being required to use a reciprocal to convert a division into a multiplication.

There could be different causes for students' problems as reported by the examiners' reports. Some of these reasons could be linked to the teaching-learning process, for example if students had not learnt certain methods or had not done sufficient application problems within a given topic. Another reason may be related to inattentiveness or stress during examination setting that may lead students to confuse terms or overlook certain requirements of the question. There could also be problems that arise due to the setting of the examination when instructions actually complicate the question as given in the last example above. It is therefore important to conduct studies to determine the cause of these problems since failure in KCSE mathematics limits advancement in studies and career choice.

Rwanda

The Rwanda Education Board, through its Examination and Accreditation Department, is responsible for the administration of public examinations in the country. Public availability of examination statistics are limited, particularly online. Examination reports are not publicly available, sample past examination papers are available in hard copy on request.

As noted in the previous chapter, Ordinary Level is sat at the end of the compulsory secondary stage, which is third year of secondary and after a total of 9 years of compulsory schooling. Table 3.5 shows the number of students who passed mathematics at O level over the years 2010–2013.

It can be seen that the uptake of the examination has increased markedly since 2010, with the pass rate almost exactly the same in 2010 and 2013. This is consistent with a net improvement in results. Further analysis is necessary to ascertain what happened in the intervening years.

It can be seen that in 2010 the number of male and female students was almost exactly the same, with a higher pass rate amongst the males. By 2013, as noted above, there has been a marked increase in the overall number of students taking the examination, it can be seen here that the increase was greater amongst female students. Whilst it continues to be the case that a greater percentage of male

Table 3.5 Uptake and pass rates for O level in Rwanda 2010–2013

Year	Number who sat for exams	Number who passed	Percentage who passed
2010	59,144	39,515	66.8
2011	77,373	47,291	61.1
2012	80,591	40,610	50.4
2013	93,484	62,544	66.9

Uworwabayeho (2014)

Table 3.6 Rwandan O level uptake and pass rate 2010–2013 by gender

Year	Female candidates			Male candidates		
	Sat	Passed	% passed	Sat	Passed	% passed
2010	29,762	18,222	61.2	29,382	21,293	72.5
2011	39,354	22,195	56.4	38,019	25,096	66.0
2012	42,841	19,663	45.9	37,750	20,947	55.5
2013	49,380	31,854	64.5	44,101	30,690	69.6

Uworwabayeho (2014)

students pass, in 2013 the absolute number of female students passing was greater than males.

Tanzania

The National Examinations Council of Tanzania (NECTA) oversees all matters related to national assessments at the end of primary and secondary (Ordinary and Advanced Level) in Tanzania. At the end of Ordinary Level, candidates sit for the Certificate of Secondary Education Examination (CSEE). All candidates must attempt the paper for Basic Mathematics.

The Basic Education Statistics of Tanzania (BEST) (Ministry of Education and Vocational Training 2013) is published in hard copy and online annually, and gives the percentage of students passing various subjects in CSEE for the previous two years including breakdown according to gender.

Achievement in CSEE

Table 3.7 gives the total number and percentage of candidates who passed Basic Mathematics at CSEE from 2009 to 2012 as given in the BEST documents (Ministry of Education and Vocational Training 2011, 2012, 2013).

Whilst the number of candidates can be seen to be increasing over this period, the pass rate has been declining steadily. Within this, the absolute number of candidates passing has been in decline from 2010 after a large increase in that year.

A breakdown by gender of CSEE results from 2009 to 2012 is given in Table 3.8.

Table 3.7 Tanzania CSEE results for basic mathematics (2009–2012)

Year of CSEE	Total number of candidates sat	Total number of candidates passed	Percentage of candidates passed
2009	240,203	42,715	17.8
2010	350,904	56,467	16.1
2011	335,960	48,886	14.6
2012	396,678	44,964	11.3

Ministry of Education and Vocational Training (2011, 2012, 2013)

Table 3.8 CSEE results for basic mathematics according to gender (2009–2012)

Year of UCE	Female candidates			Male candidates		
	Candidates sat	Candidates passed	Percentage passed	Candidates sat	Candidates passed	Percentage passed
2009	110,897	11,759	10.6	129,306	30,956	23.9
2010	160,914	16,628	10.3	189,990	39,839	21.0
2011	144,453	13,882	9.6	191,507	35,004	18.3
2012	176,655	12,349	7.0	220,023	32,615	14.8

Ministry of Education and Vocational Training (2011, 2012, 2013)

It can be seen that both the number of female candidates attempting the examination, and also the pass rate, is much lower than for male candidates.

Analysis Based on Examiners' Reports

Common errors and reasons for not doing well based on the examiners' report can be grouped in three main categories in descending order of importance.

- lack of knowledge of the topic, related properties and formulae, e.g. matrices and transformations;
- poor computational and procedural skills, e.g. solving simultaneous equations;
- difficulties with interpreting or modelling the question into a mathematical setting.

Regarding the final point, in some cases 'real life' problems contained inconsistencies which may have confused students otherwise capable of getting correct solutions. Examples are available from the authors on request.

Similarly to the analysis of KCSE examiners' reports the categories of errors can be hypothesized to arise due to reasons linked to the teaching learning process, examination stress or inattentiveness and the setting of questions. In Tanzania, language may actually have more of an influence on achievement in Mathematics because of the transition in secondary form one from Swahili to English as a medium of instruction. In a study of Form One students sampled from 21 schools across three regions general English reading comprehension and specialized mathematics vocabulary was found to be insufficient for learning Mathematics in English (Barrett, Mtana, Osaki, & Rubagumya 2014). With no special support and more than five other compulsory subjects in English in the four years it could be inferred that this problem would not decrease by the end of Form Four. Hence, it would be important to conduct studies to ascertain reasons for students' errors in CSEE and specifically how much language proficiency impacts outcomes.

Uganda

Data which can be found from Uganda comes from two main sources. Reports on the Uganda Certificate of Education (UCE) which is sat at the end of lower secondary, i.e. after 11 years of schooling, are published by the Uganda National Examination Board. In addition, since 2008, the Examination Board have published an annual National Assessment of Progress in Education which considers a sample across the country in English, mathematics and biology. The most recent available report was for 2013 (Uganda National Examinations Board 2013).

It is important to note that the policy of Universal Secondary Education, whilst not literally meaning that all Ugandan children of secondary age are in school, has meant a big expansion of students in secondary school—according to Uganda Bureau of Statistics (2011), this has gone from 728,393 in 2005 to 1,194,454 in 2009.

National Examination Data

National data is limited, what is publicly available is summarised below.

Corresponding to the policy of Universal Secondary Education, a rise in the number of candidates can be seen but a decrease in the pass rate (Table 3.9).

National data also indicates that, overall, female students are doing less well in mathematics than males (Table 3.10).

Analysis Based on Examiners' Reports

The examiners' reports go through question by question, although it is noticeable that the questions themselves are not problematized. A summary of points which were made are:

- Questions that are direct and use knowledge from one topic (sets, matrices, simultaneous equations) were well done and popular. However, adding a level of difficulty—e.g. using decimal coefficients in simultaneous equation questions-decreased the popularity and/or general performance.
- Questions on trigonometry and circle properties were generally not done well.
- There were cases of candidates using calculators either when instructed not to do so or when their use complicated the solution.

Table 3.9 Candidates and pass rates for UCE 2008–2010

Year of UCE	Total number of candidates	Percentage with distinction (1 and 2)	Percentage with credit (1–6)	Percentage with pass (1–8)
2008	197,804	2.4	31.7	82.4
2009	212,297	3.2	27.0	72.0
2010	257,943	2.1	25.4	66.0

Uganda National Examinations Board (2010, 2011)

Table 3.10 Candidates and pass rates by gender for UCE 2008–2010

Year of UCE	Total Number of candidates	Percentage with distinction (1 and 2)		Percentage with credit (1–6)		Percentage with pass (1–8)	
		Female	Male	Female	Male	Female	Male
2009	212,297	1.7	4.4	21.0	32.1	70.3	73.3
2010	257,943	1.2	2.9	20.5	29.6	62.9	68.5

Uganda National Examinations Board (2010, 2011)

The fact that students perform better on direct questions and less well on multi-step questions would appear to indicate that they do not have a deep and firm understanding of some concepts. As mentioned earlier there could be various reasons for errors. More research would be useful to provide in-depth qualitative information on the source of errors and thus provide possible interventions that would improve students' achievement in mathematics.

National Assessment of Progress in Education

Since 2008 a report has been published by the Uganda National Examinations Board on the performance of secondary form 2 students in English, mathematics and biology. This is based on a sample from across the country, for the 2013 report (Uganda National Examinations Board 2013) all 112 districts were represented, with 524 schools and 21,647 students involved. The students involved were aged between 11 to over 18, with 76 % of the students aged between 15 and 17. 45 % of the sample were girls.

The structure of the reports is much the same from year to year, although it is noted that there was a decline in performance from 2008 to 2011 with slight increases given since then (ibid: 31). Broadly consistent findings from year to year include the following.

There are large differences between different districts, corresponding to better performance in urban rather than rural areas. So, at the extremes, the mean score for sampled students in the South West area was 54.9 %, and in Mid East I it was 38.2 %.

Overall there was a large discrepancy between the performance of male and female students, with 41 % of females and 53 % of males deemed proficient in mathematics, consistent with the national findings reported above. Further, there were large differences between the performance of males and females by zone. So, the mean scores in the Near East are 42.4 % for males and 40.1 % for females, therefore a difference of 2.3, whilst in Mid North I males scored an average of 49.0 % and females 39.2 %, so a difference of 9.8 %.

Older students achieved more poorly than younger classmates, so whilst 14 year olds scored 48.3 % on average, those age 19 or more scored 39.3 %.

Discussion and Conclusions

Before considering what conclusions can be drawn from the evidence above, it is important to reflect on the nature of assessment evidence, what makes for good practice in testing and how evidence of this nature can reasonably be analysed.

It is reasonable to suppose that robust assessment data depends, in part, on 'good' tests. Black (1998) is one text which considers the nature of testing in detail, including the need for a clear sense as to what is being tested and why, the match between the taught curriculum and what is being assessed, and whether there are unintentional cultural or other biases in the question—so, for example, some children may find questions based around old people living by themselves puzzling, others may consider this entirely normal.

An analysis of test items used in the East Africa region undertaken for this study shows that one may consider that there are problems with the questions that are asked which then throws into doubt the assessment data generated by them. Examples include non-calculator techniques being tested on question requiring a calculator—so, for example, requiring the use of the reciprocal to turn a division into a multiplication in a question which required a calculator for other parts of the solution. Another example which was found was the use of an assumed average which then necessitates negative number arithmetic, so making a straightforward calculator question into a much more difficult question. Examples were also found of mistakes in questions, e.g. a literal interpretation of an algebra question giving rise to 27 year old teenagers. It is noticeable that the examiners' reports, as considered above, give advice to students and teachers but do not ever, as far as could be found, acknowledge that there might have been a problem with the original question itself.

Beyond this there are problems in the reporting. It was not always possible to match assessment data with the questions which gave rise to the assessment data. Whilst from an examination security point of view this may make sense, particularly if papers are to be reused, this does mean that the analysis is conducted with limited information available.

From the complex account of attainment contained in the discussion above, a number of key themes can be seen to be emerging.

Comparing assessment data across the region is complicated by different school systems, intended age of candidates, grading systems and reporting protocols. In working towards harmonization, greater commonality in these respects would, in principle, facilitate the networking of resources in achieving economy of scale in promoting improvement in access, quality and equity across the region.

As things stand, pass rates and average scores overall are low. Irrespective of the quality of the examination process, this clearly demonstrates a mismatch between the intentions of the policy makers and the reality of student performance across the region. Possible causes to investigate include ensuring that the curriculum is relevant and motivating to students, with clear progression in topics. Consistent with the paper on teacher education in this volume the quality of teaching may well also be a factor. The robustness of the examination systems and the consistency of test items needs also to be investigated. In all countries English is the medium of instruction in the secondary sector but only exceptionally used by students outside the classroom. Difficulties reported in students answering word problems need to be analysed in this light.

A related issue is that of difficulties in students attempting multi-step questions. One might consider that language difficulties, alongside lack of total security in learning, mean that the perception of difficulty in multi rather than single step problems increases disproportionately. Further investigation is required as to precisely what the underlying issues are, in looking to support high quality secure learning which students can apply in English across a range of simple and more complex contexts.

Male students in general are taking mathematics in larger numbers and achieving higher pass rates than females. However, this is not consistently the case. In Rwanda there are more female candidates in mathematics, albeit with a lower pass rate. And the Uwezo figures for primary children showed higher attainment for girls. Further, where regional data was available, particularly in the Ugandan National Assessment of Progress data, complex patterns confounding gender and region can be seen. These findings give reason to believe that there is considerable potential for male and particularly female students across the region to be achieving higher results than currently, with both general and gender specific interventions needed.

Beyond gender differences, it is very noticeable that there are differences by region, particularly with students in urban areas achieving more highly than their rural counterparts. Again, this gives reason to believe that there is substantial potential for improved standards across the region and particularly in rural areas.

It has been noted that student numbers in secondary schools across the region have been increasing over the reporting period. It can reasonably be assumed that, in general, students attending school now who would not have done so previously will be achieving on average at a lower level than others. Means of ensuring a motivating, progressive enjoyable curriculum for all learners, not just the highest attainers, need to be evaluated in ensuring school systems fit for the 21st century.

References

Barrett, A. M., Mtana, N., Osaki, K., & Rubagumya, C. (2014). *Baseline study report, 'Strengthening secondary education in practice': Language supportive teaching and textbooks in Tanzania (LSTT)*. Bristol, UK: University of Bristol.

Black, P. J. (1998). *Testing: Friend or foe? Theory and practice of assessment and testing*. London: The Falmer Press.

Hungi, N., Makuwa, D., Ross, K., Saito, M., Dolata, S., van Cappelle, F., & Vellien, J. (2010). *SACMEQ III project results: Pupil achievement levels in reading and mathematics*. Praetoria: SACMEQ.

Kenya National Examination Council. (2012). *The year 2011 KCSE examination report*. Nairobi: KNEC.

Kenya National Examination Council. (2013). *The year 2012 KCSE examination report*. Nairobi: KNEC.

Ministry of Education and Vocational Training. (2011). *Basic education statistics in Tanzania (BEST) 2007–2011: Revised national data*. Dar es Salaam: MOEVT.

Ministry of Education and Vocational Training. (2012). *Basic education statistics in Tanzania (BEST) 2008–2012: Revised national data*. Dar es Salaam: MOEVT.

Ministry of Education and Vocational Training. (2013). *Basic education statistics in Tanzania (BEST) 2009–2013: Revised national data*. Dar es Salaam: MOEVT.

Uganda Bureau of Statistics. (2011, October 12). *Trend of key secondary school education indicators (2005–2009)*. Retrieved on November 29, 2013, from http://www.ubos.org/?st=pagerelations2&id=21&p=related%20pages%202:Education%20Statistics

Uganda National Examinations Board. (2010). *Report on work of candidates UCE 2009.* Kampala: UNEB.

Uganda National Examinations Board. (2011). *Report on work of candidates UCE 2010.* Kampala: UNEB.

Uganda National Examinations Board. (2013). *The achievement of S2 students and teachers in English language, mathematics and biology*. Kampala: Uganda National Examinations Board.

Uwezo East Africa. (2012). *Are our children learning? Literacy and numeracy across East Africa.* Dar es Salaam: Uwezo East Africa.

Uworwabayeho, A. (2014). *Rwandan S3 examination results.* Private communication, August 28, 2014.

Chapter 4
Teaching and Learning Mathematics: Insights from Classrooms in East Africa

Anjum Halai

Abstract This chapter looks at issues of teaching and learning mathematics in under resourced classrooms in low-income, post-colonial East African countries. It provides insights into issues arising from externally driven pedagogic reforms introduced to promote learners' active participation in mathematics classrooms without critical engagement with the wider socio-cultural norms within which the reform was introduced. The chapter further raises issues arising due to the language of instruction policies that mandated learning mathematics in English (language of the ex-colonisers), so that learners resorted to 'safe talk' and did not necessarily engage with mathematics. Finally, the chapter highlights issues for participation of boys and girls in mathematics due to traditional and stereotypical understanding of gender roles in society. The chapter ends with recommendations for policy and practice to improve teaching and learning in traditional mathematics classrooms in low-income countries.

Introduction

Mathematics teaching in an increasingly global and technological world of the 21st century must go beyond the traditional delivery and transmission of content knowledge premised on the learner as tabula rasa and passive recipient of knowledge. It would need to promote learners' active participation in the classroom to learn conceptually sound mathematics, develop mathematical thinking skills including communication of mathematical ideas, generating and interpreting data and problem solving (UNESCO 2014; World Bank 2007, 2011). This view of teaching and learning is a major shift from the traditional perspective as described above and raises several

A. Halai (✉)
Aga Khan University Institute for Educational Development East Africa,
Salaama House Urambo Street, PO125, Dar es Salaam, Tanzania
e-mail: anjum.halai@aku.edu

A. Halai and G. Tennant (eds.), *Mathematics Education in East Africa*,
https://doi.org/10.1007/978-3-319-27258-0_4

issues for classrooms in the less technologically developed often traditional, post-colonial, social and cultural contexts such as in East Africa. For example, a major issue would be transforming pedagogic practice in the classrooms to support learners' active participation. An active role of learners would necessarily require challenging assumptions about the role and status of the learner in relation to the teacher. Likewise, the colonial heritage is often reflected in the classrooms through the imposition of a language of instruction (LoI) that is usually not the first or the second language of the learners. There is ample evidence that patterns of interactions in classrooms where teaching is through a third language of the learners are not conducive to learning mathematics. Finally, traditional societies often have highly gender specific roles for boys and girls in the society. These societal perceptions of gender roles have major implications for learners' equitable participation in mathematics classrooms as curriculum materials and teaching often show an implicit gender bias.

This chapter aims to provide an understanding of the processes of teaching and learning and related issues in mathematics classrooms in the sub-Saharan Africa with a focus on Kenya, Rwanda, Tanzania and Uganda. Mainly the issues discussed are those noted above. It is recognised that teacher quality and curriculum are also significant issues that impact the quality of teaching and learning in classrooms. Issues related to teacher quality are covered mainly by Kajoro in this volume, while Karuku and Tennant (in this volume) have discussed issues related curriculum and syllabus in mathematics.

Teaching Practice in Mathematics Classrooms: A Reform Agenda

Classrooms in sub-Saharan Africa are often over-crowded, under-resourced, with teacher centered patterns of interaction that provide little scope for learners to discuss ideas through discussion with each other or with the teacher. For example, Hardman, Abd-Kadir, Agg, Migwi, Ndambuku, & Smith (2009) profiled patterns of interaction in public schools in sub-Saharan Africa and maintained that:

> Overall, the classroom discourse in sub-Saharan African classrooms was found to be highly ritualised creating a semblance of curriculum coverage, knowledge and understanding. The most prevalent methods of teaching were teacher explanation punctuated by a question and answer approach, choral responses, pupils copying from the chalkboard; written exercises and teachers marking pupil work. Teacher-led recitation, with its constant demand for pupil participation, ensured teachers maintained tight control of the learning environment. (p. 65–86)

Specifically with reference to mathematics classrooms in public schools in Tanzania, Kajoro maintains that teaching practice in the classroom is characterised by teaching of algorithms, procedures and facts that are rote memorized and reproduced by the learner, classroom teaching is mainly teacher led and informed by theoretical perspectives that position the teacher in the role of a transmitter and the student in that of a passive recipient (Kajoro 2012, 2016).

Largely in response to issues noted above, such as the rote nature of learning mathematics and the persisting evidence of students' underachievement in mathematics (as noted in Tennant and Sarungi this volume), there was large-scale reform in mathematics education in the East Africa region. A key feature of these reform projects was an emphasis on introducing more learner centred approaches to teaching and learning especially in science and mathematics with teaching and learning strategies such as problem solving, project work, cooperative learning, and activity based learning. For example, In Tanzania the large scale Secondary Education Development Programme II[1] (SEDP II) had an explicit focus on improving the quality of teaching and learning especially in science and mathematics classrooms through "ensuring that student-centered methodologies/pedagogy are reinforced in the classroom" (MoEVT 2010: viii).

In Kenya the Strengthening Science and Mathematics in Secondary Education (SMASSE[2]) project is by far the most significant initiative in science and mathematics education in public sector schools and in-service teacher education (JICA 2007). Discussing issues that the SMASSE had tried to address Wambui, a national trainer under this project holds that "teachers have been found to present lessons that are too much teacher-centred with the teacher as the main actor and sometimes the only actor in the classroom as students remain passive recipients. [--] a student-centred lesson should be enhanced from two complementary elements; (i) placing more responsibility in the hands of students, and (ii) requiring the teacher to serve as a mentor and facilitator in presenting knowledge especially to students and fellow teachers in the teaching/learning process" (Wambui 2005: 214).

Along similar lines as in Kenya a significant initiative in improving science and mathematics teaching in secondary schools in Uganda was the project the Secondary Science Education and Mathematics Teachers (SESMAT) (MoES 2007). In a critical examination of the SESMAT, it is maintained by Ssebbunga-Masembe, Bisaso, Kyasanku, Nakawuki, & Nakabugo (2013: 158) that, "SESMAT is premised on the assumption that teaching of Science should be made practical and teachers should be transformed. It provides in service training to science teachers to equip, retrain and develop competencies in them. SESMAT is also based on the paradigm shift from the conventional/traditional talk and chalk to a more practical and learner centered approach to teaching science and mathematics."

A common thread running through the reform and improvement in school mathematics teaching was a shift away from teacher centered and "chalk and talk" approaches towards an activity based, constructivist or similar teaching strategies.

[1]The Secondary Education Development Programme II (SEDP II) over 2010–2015, was a continuation of SEDP I, which was implemented between 2004 and 2009, building on the national goals of secondary education provision. (http://www.pmoralg.go.tz).

[2]SMASSE programme is a bilateral technical cooperation intervention between Japan, (JICA), and Kenya (MOE) (1998–2003 pilot; 2003–2008 national programme). The goal of SMASE was to enhance the capability of learners in Mathematics and Science through strengthened teaching and learning.

A sustained challenge is that these "student centered" pedagogic practices and teaching strategies remained at best at a superficial level so that issues in student learning and outcomes persisted (Kamau, Wilson, & Thinguri 2014; Ssebbunga-Masembe et al. 2013; Wambui 2005). For example, on the basis of evidence generated from their annual learning assessment conducted in Kenya, Uganda and Tanzania, UWEZO a regional think tank holds that "Less than one third of the pupils enrolled in standard three in East Africa possess the basic numeracy or literacy skills" (UWEZO 2014: 7). Of course teaching strategies and pedagogic practice are one element of the quality inputs, but nonetheless a crucial element and their appropriateness and relevance need to be critically considered.

Models of teaching practice that are "student centered" are usually derived from knowledge emerging mainly from the global North or West (Halai 2011a; Tabulawa 2013). An issue is that introducing teaching strategies and pedagogic approaches "imported through reform" may not necessarily be appropriate to the socio-cultural setting within which African teachers and learners operate. For example, Vavrus (2009: 303) argues on the basis of an ethnographic study of teacher education practice in Tanzania that "the cultural, economic, and political dimension of teachers' practice need to be considered alongside efforts to reform the country's education system." Tabulawa (2013) has looked in depth at the issue of pedagogical reforms in sub-Saharan Africa and maintains that teachers (and parents) view quality in education mainly as high achievement in examination and tests. Hence, "Pedagogical innovations whose utility in this regard is not obvious to teachers and students are unlikely to be embraced" (p. 18). Likewise, Barrett (2007: 290) points out the futility of dualisms such as "child centred" versus "teacher centred" practice, saying that teachers incorporate notions of "progressive practice" into traditional notions of good pedagogic practice.

To conclude from the above, improving the quality of teaching practice in mathematics classrooms is urgently required. However, it is not simply a function of change in teaching methods and techniques. A sustainable change would require questioning the assumptions that underpin practice and introducing more socio-culturally embedded approaches to teaching.

Language of Instruction in Teaching and Learning Interactions: Cognition and Communication

A noteworthy feature of the teaching and learning processes in mathematics is the impact of the language in education policy in the region. In the education system in postcolonial countries the language of instruction (LoI) is usually the language of the former colonial power because of its perceived or real benefits contributing to the economic development and upwards social mobility of the learners (Jones 2013; Brock-Utne 2012). For example, in, Kenya, Uganda, Tanzania and now also Rwanda the language of instruction in secondary and post secondary education is English. A quick over view of the policy of LoI in East Africa is provided in Table 4.1.

Table 4.1 Language of instruction policy in East Africa

Country	Level of education	Language of instruction
Tanzania (mainland) Source: Government of Tanzania (2011)	Pre-primary (two years)	· Kiswahili
	Primary standard I–VII	· Kiswahili
	Secondary form I–VI and post secondary	· English[a]
Zanzibar (a semi-autonomous region within Tanzania)	Pre-primary (two years)	· Kiswahili
	Primary education standard I–VII	· Kiswahili · Starting 2010, English is the LoI in mathematics and science
	Secondary form I–VI and post secondary education	· English
Kenya Source: Kenya Institute of Education (2006) and USAID (2012)	Pre-primary (two years) Early primary education (1–3 grades)	· Mother tongue · Kiswahili or English in the urban schools or in 'linguistically heterogeneous' areas
	Upper primary	· English
	Secondary and post secondary education	· English
Uganda Source: National Curriculum Development Centre Uganda (2008, p. 3)	Pre-primary Early primary (grades 1–3)	· Mother tongue or language of the community · English in 'linguistically heterogeneous' areas
	Primary	· English
	Secondary and post secondary education	· English
Rwanda Source: Republic of Rwanda Ministry of Education (2010, p. 14)	All levels of education	· English[b]

[a]The new education policy launched in March 2015 aims to introduce Kiswahili as the LoI at all levels of education (http://www.moe.go.tz)
[b]In 2008, there was a policy change and French language was dropped as a LoI

From Table 4.1, it is seen that in public schools in the East Africa, beyond primary education level, mathematics is taught in English, a language that is a second or third language of the learners and the teachers. There are consequences of this policy decision on the nature of classroom interactions and the quality of teaching and learning in mathematics (Halai and Karuku 2013). Certain specific features are noted in the patterns of interaction when mathematics teaching and learning takes place in a second or third language as the LoI.

'Safe Talk' is a significant concern in mathematics classrooms where teachers and learners revert to certain 'safe' approaches that give a semblance of participation, such as learners providing brief one-word choral responses to teachers' questions, repeating the information provided by the teacher, and the teacher progressing without probing for individual understanding or seeking new information (Chick 1996; Brock-Utne 2005). These approaches are called 'safe talk' because they help learners save face by not providing wrong answers publicly, and they protect the teacher by allowing the lesson to progress along pre-determined lines, by avoiding unexpected issues arising due to the teacher releasing control to allow learners participation (Chick 1996; Brock-Utne 2005; Martin 2005). A concern with 'safe talk' is that it might give the appearance of the lesson progressing but does not necessarily provide an opportunity to the learners to engage in conceptual learning in mathematics.

For example, provided below is an example of 'safe talk'. This is a lesson extract from a public secondary school in Zanzibar. Learners were introduced to the graphical method of seeking a solution to quadratic equations, a topic that is often considered quite difficult by learners and teachers, and is a significant building block for learners' progress to advanced mathematics including functions. While the language of instruction is English, the classroom interactions in the extract below are in English and Kiswahili, an issue that is discussed later in this section.

Box: 1

1. T: **Thanks very much. We said that there are four methods. We have four method to solve quadratic equations Today this is the last method. How many are they?**

2. Class: nne

(Class: four)

3. T: Kwa hivo leo tiutazungumza method ya mwisho ya kusolve quadratic equation ya graphical.... Asema the general quadratic general equation who can try to tell us the general equation

(T: So, today we are going to discuss the last method used to solve quadratic equation graphica... the general quadratic equation, who can try to tell us the general equation, general equation of quadratic equation. Who can try general equation ?)

4. Pl: ax square plus bx plus c is equal to zero

5. T: **(repeats what the student was saying as he wrote it on the board) ax square plus bx plus c is equal to zero. Thank you very much this can be solved graphical (teacher writes on the board as he speaks).**

6. P: graph

7. T: Kama ulivoweza kuisolve kwa kutumia capitalization kama ambavyo uliweza kuisolve kw akutumia fomula vile vile una uwezo wakusolve kwa kutumia graphical sawaa lakini jambo la mwanzo ufahamu kitu kukisolve au kukitengeneza kama hujajua namna ya kukitengeneza si ndio. Utatengeza wali wakati hujui hata namna ya kutengeza wali hata ukapewa vifaa huwezi kuupika mpaka ujue si ndio.

(T: **As you had solved it by factorization, As you had solved it by using formula, you can also solve it by using graphical. Ok but the first thing you must know is how to draw the graph because you wont be able to solve an equation or prepare something which you don't know how it**

is worked out or prepared. How could you make rice if you don't know how it is made even if you are given the ingredients. You cant make it if you don't know how to right?)
8. Class: Ndio
(Class: **right)**
9. T: Au sio (student's name)
(T: **Am I right (student's name)**
10. P: Ndio
(P: Sure)
11. T: Haya sasa namna ya kulitengeneza grafu. Vipi namna ya kulichora graph. How can we draw the graph sawa
(T: Ok now how to draw the graph, the graph, how to draw the graph, how can we draw the graph. How you can draw graph okay.)
12. P: Ndio
(P: Okay)
13. T: haya unapopewa equation kama hii general kwamba ax square plus bx plus c is equal to zero [inaudible] graph tulipokuwa tukitafuta factor tulitafuta factor ya bx au sio
(T: Right when you are given an equation, which reads like this general factor right? Ax square plus bx plus c is equal to zero.[inaudible] when we were solving using factor method we tried to find the factor of bx is that right?)
14. Class: eeee
(Class: Yes)
15 T: Tulipokuwa tukitafuta kwa fomula tukiita a b c si ndio? tulipokuwa tukitafuta kwa njia hii nyingine (turning to students for the answer)
(T: When we were solving using formula method we called it a b c right? when we were solving it using this other method (turning to students for the answer)
16 P: complete square
(P: complete Square)
17. T: Complete square plus c subject si ndio
(T: Complete square plus c subject right?)
18. P: eee
(. P: yees)
Source: SPINE Project (Student Performance in National Examination) http://www.bristol.ac.uk/spine/

The data extract above was symptomatic of 'safe talk'. The teacher's contributions were lengthy and learners' responses were brief and usually in one word so that it was difficult to assess the learning that might have taken place. Moreover, the responses being choral in nature did not differentiate between learners who actually understood what the teacher was doing and saying from those who did not. The teacher moved on to the next steps in the lesson, and did not attempt to engage the learners on the basis of their brief responses. The lesson thus appeared to have progressed with little evidence of the nature of mathematics learning.

Code switching and translation are two other features of language use in the course of teaching and learning through a second or third language as an LoI. A significant implication of the use of code switching and translation in the course of teaching and learning is that it provides a scaffold to the learners and the teachers who are not proficient in the LoI (Halai 2009). However, the policy of language in education takes an 'ideal' perspective of language use in the classroom and does not recognise or support the use of multiple languages in the classroom (Jones 2013). From this perspective all teaching and learning interactions are expected to take place only in the LoI. Hence, high stakes school leaving examinations in Kenya, Rwanda, Tanzania and Uganda are conducted entirely in English. A consequence is that learners are not able to perform well and evidence their learning in school leaving examinations. For example, on the basis of a large scale project looking at students achievement in national Form Two examination in mathematics, biology and English, as noted by Rea-Dickins, Khamis, & Olivero (2013) that "making linguistic modifications in the statement of examination items had a significant impact on the way in which the learners processed the item mathematically. In one instance, on the original item, there were no correct answers, with 30 % of the learners not even attempting to answer the items. By way of contrast, all modified tasks achieved some correct or partially correct answers, with only a very small percentage of learners failing to answer the question at all" (Rea-Dickins et al.: 111). The report also notes that students who were able to provide a correct translation of the question statement performed better on the test items.

In sum a consequence of use of a second or third language as LoI is that mathematics learning remains at a superficial level. For effective teaching with improved learning outcomes, the policy of introducing English as a language of instruction needs to question deep-rooted assumptions about the role of language in learning. Policy assumes use of only one 'official LoI' and does not support use of multiple linguistic devices in the course of teaching and learning, whereas language in practice is fluid and draws from all linguistic resources of the learners. A nuanced interpretation and implementation of the LoI could mean that learners' languages are seen as a resource that the teachers could employ to facilitate the participation of learners in the process of learning. Finally implementation of the LoI must necessarily be seen in conjunction with policies and practices that look into examination and teacher education to ensure that teaching and examination processes also value the learners' language as a resource in the classroom.

Boys' and Girls' Participation in Learning Mathematics: Gender Equity Issues

Sub-Saharan Africa has shown the fastest growth in gross enrolment to lower secondary schools where enrolment more than doubled reaching 49 % in 2011, with the fastest growth in gender parity in education (UNESCO 2014). Rwanda is widely recognised as leading the continent towards gender parity in girls' access to both primary and secondary schools (Pro-Femmes Twese Hamwe and VSO 2013). However, there are persistent and often wide inequities in completion and access. Girls' access to post primary education in Africa is impacted due to several social and cultural reasons including concerns for girls' safety in case a secondary school is not available in close vicinity.

Achievement in mathematics also reflects gender bias in favour of boys. For example, Saito (2011) looked at the SACMEQ results to consider the trends in gender equality in learning achievement in Southern and Eastern Africa and maintains, "In Malawi, Kenya, Tanzania Mainland, and Zambia, boys performed significantly better in all of the mathematics domains" (p. 11). Gender disparity in achievement is well documented in Tennant and Sarungi (this volume). This persisting disparity raises questions for the quality of participation in mathematics for boys and girls.

Beyond access, girls' and boys' participation, especially in science and mathematics classrooms is qualitatively different. Factors that contribute to gender inequity in teaching and learning processes and outcomes in mathematics classrooms in sub-Saharan Africa include: highly gendered identities, roles and responsibilities in the community with girls seen mainly as home-makers and boys as career makers (Chege 2006; Mirembe and Davies 2001; Pro-Femmes Twese Hamwe and VSO 2013); lack of time for girls to do homework due to household chores, non-referencing of female characters in mathematics textbooks (Githua 2013; Githua and Mwangi 2003); teachers' and parents' higher expectation from boys as compared to girls in mathematics achievement (Mlama, Dioum, Makoye, Murage, Wagah, & Washika 2005); teachers' and students' perceptions of mathematics as a masculine discipline not suitable for girls (Halai 2011b).

To address issues of girls' and boys' inequitable participation in mathematics classrooms, FAWE initiated an intervention entitled 'Science Mathematics and Technology' that included a gender responsive pedagogy to address the quality of teaching in African schools including those in the East African region (http://www.fawe.org[3]). They maintained that the innovative pedagogical approaches (e.g. group discussions, demonstration) were not necessarily inherently gender inclusive and that these approaches had to be seen critically for their implications for boys and girls and recommend that teacher education must prepare teachers who

[3]Forum for African Women Educationalists (FAWE) is a significant pan-African Non-Governmental Organisation working in 32 African countries to empower girls and women through gender-responsive education. For more information see http://www.fawe.org.

are well versed in a gender responsive pedagogy (Mlama et al. 2005). Similarly, there are a number of initiatives in improving gender parity in education especially in science and mathematics teaching. For example, the SMASSE and SESEMAT projects noted earlier in the chapter have explicit mention of promoting gender parity in access and participation. Likewise national 'vision statements' of the countries in the region aim to ensure that gender mainstreaming in society is achieved through education and other national or regional policy commitments (SID 2013). However, the extent to which the policy commitments in education translate into gender inclusive processes in the classrooms remains unclear. Pro-Femmes Twese Hamwe and VSO (2013) make several considered recommendations on the basis of a systematic study to look at gender equality in teaching and educational management in Rwanda. They maintain that strong affirmative action is required to support girls' participation in education and that the scope of action must go beyond the confines of the school to take account of teacher education and community participation. In brief, further research and evaluation studies are required to look at the implementation of the policies and initiatives for gender equity at grassroots level in the education.

Concluding Reflections

From the foregoing discussion it may be deduced that the processes of teaching and learning take place in classrooms but the culture of the classroom does not exist in a microcosm, rather it reflects the values and beliefs of the wider community within which classrooms are situated. Hence, societal values that put a premium on English as compared to Kiswahili are reflected in the classroom dynamics that also appear to value English as the LoI. Similarly, beliefs about the gendered roles and responsibilities of boys and girls are reflected in the differentiated expectations in mathematics that teachers have from boys and girls as learners.

To conclude, prevailing teaching and learning practices in the mathematics classrooms in East Africa need to change to enable deeper and greater engagement of learners and a shift from a highly content driven to skills based teaching and learning. However, introducing change through notions of "good practice" drawn from mainly Euro-western social and cultural contexts needs to be problematized. For improved learning outcomes in mathematics in the East African classrooms more nuanced and contextually appropriate models of teaching and learning must be developed.

References

Barrett, A. M. (2007). Beyond the polarisation of pedagogy: Models of classroom practice in Tanzanian primary schools. *Comparative Education, 43*(2), 273–294.

Brock-Utne, B. (2005). Language-in-education policies and practices in Africa with a special focus on Tanzania and South Africa: Insights from research in progress. In J. I. Zajda (Ed.),

International handbook on globalisation, education and policy research (pp. 549–565). Dordrecht, The Netherlands: Springer.

Brock-Utne, B. (2012). Language policy and science: Could some African countries learn from some Asian countries? *International Review Education, 58*, 481–503.

Chege, F. (2006). Teachers' gendered identities, pedagogy and HIV aids education in African settings within the ESAR. *Journal of Education, 38*, 26–44.

Chick, J. K. (1996). Safe talk: Collusion in apartheid education. In H. Coleman (Ed.), *Society and the language classroom* (pp. 21–39). Cambridge, UK: Cambridge University Press.

Githua, B. N. (2013). Secondary school students' perception of mathematics formative evaluation and the perceptions' relationship to their motivation to learn the subject by gender in Nairobi and Rift Valley provinces in Kenya. *Asian Journal of Social Science and Humanities, 2*(1), 174–183.

Githua, B. N., & Mwangi, J. M. (2003). Students' mathematics self concept and motivation to learn mathematics: Relationship and gender differences among Kenya's secondary school students in Nairobi and Rift Valley provinces. *International Journal of Educational Development, 23*, 487–499.

Government of Tanzania. (2011). *Education.* Retrieved March, 2012 from www.tanzania.go.tz/educationf.html.

Halai, A. (2009). Politics and practice of learning mathematics in multilingual classrooms: Issues for policy and practice. In R. Barwell (Ed.), *Multilingualism in mathematics classrooms: Global perspectives* (pp. 47–62). London: Multilingual Matters.

Halai, A. (2011a). Researching education research methodologies in the "South": An introductory discussion. In A. Halai & D. Wiliam (Eds.), *Research methodologies in the South* (pp. 1–18). Karachi: Oxford University Press.

Halai, A. (2011b). Equality or equity: Gender awareness issues in secondary schools in Pakistan. *International Journal of Educational Development, 3*(1), 44–49.

Halai, A., & Karuku, S. (2013). Implementing language-in-education policies in multilingual mathematics classrooms: Pedagogical implications. *Eurasia Journal of Mathematics, Science and Technology Education, 9*(1), 167–176.

Hardman, F., Abd-Kadir, J., Agg, C., Migwi, J., Ndambuku, J., & Smith, F. (2009). Changing pedagogical practice in Kenyan primary schools: The impact of school-based training. *Comparative Education, 45*(1), 65–86.

JICA. (2007). *Strengthening of mathematics and science in secondary education project in Kenya: A capacity development analysis.* Nairobi: Japan International Cooperation Agency.

Jones, J. M. (2013). The 'ideal' vs. the 'reality': Medium of instruction policy and implementation in different class levels in a western Kenyan school. *Current Issues in Language Planning, 15*(1), 22–38.

Kajoro, P. (2012). Conventionalization of numerals in Tanzania. *Mathematics Teaching, 227*, 15–22.

Kajoro, P. (2016). Transition of the medium of instruction from English to Kiswahili in Tanzanian primary schools: Challenges from the mathematics classrooms. In A. Halai & P. Clarkson (Eds.), *Teaching and learning mathematics in multilingual classrooms: Issues for policy, practice and teacher education* (pp. 73–88). Rotterdam: SENSE Publishers.

Kamau, K. J., Wilson, K. L., & Thinguri, R. (2014). An evaluation of effectiveness of SMASSE program in performance in science and mathematics in primary schools in Kenya. *International Journal of Educational Research, 2*(6), 1–10.

Kenya Institute of Education. (2006). *Primary education mother tongue handbook.* Nairobi: Ministry of Education.

Martin, P. (2005). 'Safe' language practices in two rural schools in Malaysia: Tensions between policy and practice. In A. M. Y. Lin & P. Martin (Eds.), *Decolonisation, globalisation: Language in education policy and practice* (pp. 74–97). Clevedon: Multilingual Matters.

Mirembe, R., & Davies, L. (2001). Is schooling a risk? Gender, power relations, and school culture in Uganda. *Gender and Education, 13*(4), 401–416.

Mlama, P., Dioum, M., Makoye, H., Murage, L., Wagah, M., & Washika, R. (2005). *Gender responsive pedagogy: A teacher's handbook*. Nairobi: FAWE.

MoES. (2007). *Project proposal for the expansion of the secondary science education and mathematics teachers' (SESEMAT) pilot project; support towards implementation of universal post-primary education and training (UPPET)*. Kampala: Ministry of Education and Sports.

MOEVT. (2010). *Secondary education development plan II*. Dar es Salaam: MoEVT.

National Curriculum Development Centre, Uganda. (2008). *Primary school curriculum: Primary 3*. Kampala: Ministry of Education and Sports.

Pro-Femmes Twese Hamwe, & VSO. (2013). Gender equality in teaching and educational management: A summary report of research undertaken in Rwanda. Kigali: VSO Rwanda.

Rea-Dickins, P., Khamis, Z., & Olivero, F. (2013). Does English-medium instruction and examining lead to social and economic advantage? Promises and threats: A Sub-Saharan case study. In E. Erling & P. Seargeant (Eds.), *English and international development* (pp. 111–140). Avon: Multilingual Matters Ltd.

Republic of Rwanda, Ministry of Education. (July 2010). *Education Sector Strategic Plan: 2010-2015*. Retrieved from https://www.unicef.org/rwanda/RWA_resources_educssp2010.pdf

Saito, M. (2011). *Gender equity in education: Looking beyond parity*. Paris: UNESCO.

SID. (2013). *Tracking gender equality commitments in the East African Community: A monitoring and evaluation framework*. Nairobi: Society for International Development.

Ssebbunga-Masembe, C., Bisaso, R., Kyasanku, C., Nakawuki, R. C., & Nakabugo, G. M. (2013). Examination of locally and externally-initiated teacher professional development (TPD) programmes for science and mathematics teachers in Ugandan secondary schools. *Journal of International Cooperation in Education, 15*(3), 149–168.

Tabulawa, R. (2013). *Teaching and learning in context: Why pedagogical reforms fail in sub-Saharan Africa*. Dakar, Senegal: CODESRIA.

UNESCO. (2014). *Teaching and learning: Achieving quality for all. EFA global monitoring report*. Paris: UNESCO.

USAID (2012). Primary Math and Reading Program (PRIMR) Kenya, Education Policy Study Report. Retrieved from http://www.pdf.usaid.gov/pdf_docs/pdacy060.pdf

UWEZO. (2014). *Are our children learning? Literacy and numeracy across East Africa 2013*. Nairobi: Twaweza.

Vavrus, F. (2009). The cultural politics of constructivist pedagogies: Teacher education reform in the United Republic of Tanzania. *International Journal of Educational Development, 29*(3), 303–311.

Wambui, N. (2005). *Study on mathematical achievement using the climbing learning method in Kenyan secondary school*. Paper presented at the Mathematics Education into the 21st Century Project Universiti Teknologi Malaysia Reform, Revolution and Paradigm Shifts in Mathematics Education. Malaysia: Johor Bahru.

World Bank. (2007). *Developing science, mathematics, and ICT education in Sub Saharan Africa: Patterns and promising practices. Working paper no. 101*. Washington, D.C.: World Bank.

World Bank. (2011). *Learning for all: Investing in people's knowledge and skills to promote development. Education Strategy 2020*. Washington, D.C.: World Bank.

Chapter 5
Mathematics Teacher Training in East Africa

Peter Kajoro

Abstract This chapter firstly briefly considers the history of teacher training worldwide. It then goes on to specifically look into the status of mathematics teacher training in East Africa by reviewing the pathways to becoming a mathematics teacher; the entry criteria and guidelines; the structure and curriculum in mathematics teacher preparation; the profile of tutors who prepare the mathematics teachers; and finally, the professional development programmes for both mathematics teachers and tutors. The chapter concludes by highlighting some differences in a number of aspects regarding the mathematics teacher training and urges educational stakeholders to speedily embark on harmonizing examination grading systems and the structure and curriculum of mathematics teacher education. It further calls on alignment of teachers' and tutors' professional development. Harmonization of various elements of mathematics teacher training is seen as a way of ensuring that all East Africans have equitable opportunities on the labour market under the East African common market protocol.

Introduction: The Nature of Teacher Training

Whilst it might appear to be self-evident that teachers need to be trained before starting to work in the classroom, this is in fact quite a new idea in the broad sweep of history. According to the Institute of Education (2014), formal teacher training started in the UK in 1846, whilst in South Africa the first teacher training college was established in 1838 (Wolhuter 2006). The assumption before this time, with formal education being only for a minority, was that subject knowledge was sufficient, without the need also to engage with how to go about teaching.

P. Kajoro (✉)
Aga Khan University Institute for Educational Development East Africa,
Salaama House Urambo Street, 125, Dar es Salaam, Tanzania
e-mail: peter.kajoro@aku.edu

A. Halai and G. Tennant (eds.), *Mathematics Education in East Africa*,
https://doi.org/10.1007/978-3-319-27258-0_5

The idea that teachers need to be trained has now gained wide acceptance around the world, so that in an American context looking at grade 4 and 5 mathematics results, "… we find certified teachers consistently produce stronger student achievement gains than do uncertified." (Darling-Hammond, Holtzman, Gatlin, & Vasquez Heilig 2005: 2). The increasing need for teachers to engage with teaching methods beyond those they experienced themselves as children in school (see Chaps. 4 and 6 in this volume considering interactive teaching methods and ICT respectively) adds weight to the argument that training is needed before teachers start to teach (or possibly, alongside their first few years of teaching).

A range of issues arise in the discourse around teacher training. One key issue is encapsulated in the phrase 'apprenticeship of observation' (Lortie 1975). The point here is that, by the time candidates for teaching start their training, they have almost always spent a very large amount of time in the classroom as children observing teachers at work, far more than any other profession. This means that beginning teachers bring with them deeply embedded models as to what teaching is supposed to look like, which are then very difficult to change. Similarly, teacher trainers have also spent many hours as children with teachers. Whilst they may have spent considerable time considering the nature of effective teaching, they also may be struggling to break down pre-existing models, with the very possible result that they end up teaching, e.g. interactive teaching methods by lecturing.

How teacher-training courses are organized has also attracted a certain amount of attention across the world. The extent to which courses are based in school rather than College or University, the extent to which training is front-loaded or alongside the initial teaching, has been varied in different models across the world—in a USA context see Constantine, Player, Silva, Hallgren, Grider, & Deke (2009), and in a UK context see Moses (2009). Allied to this, the extent to which beginning teachers require a theoretical understanding rather than the chance to watch 'good' teachers in operation and the chance to practice themselves is also up for debate (in a UK context see, for example, SCETT 2011).

All of this needs to be put against a background of teacher supply. Unless class sizes are made arbitrarily high, the moves towards Universal Primary and indeed Universal Secondary Education necessarily increases the number of teachers required. But this in turn raises many issues regarding how popular the teaching profession is with members of any given society; to be able to recruit the big number of teachers required, the teaching profession must be acceptable to a large number of suitably qualified people. Strongly related issues then become the extent to which teachers feel valued in society, and whether teaching is regarded as a viable choice for those for whom other professions are a possibility.

At a policy level, the four East African countries have underscored the importance of both mathematics and mathematics teaching and learning for their respective national economic growth and social development in a number of their documents. The national development vision documents in particular (The United Republic of Tanzania 1999; Republic of Rwanda, Ministry of Finance and Economic Planning 2000; Government of the Republic of Kenya 2012; Government of Uganda 2013) clearly articulate how quality mathematics teaching and learning will play a significant role in enabling them realize their visions of becoming middle-income economies.

In accordance with the discussions above, moves to improve the quality of mathematics learning have therefore necessarily included consideration of the quality of teacher training. It is in the light of the importance of teacher education in the context of mathematics teaching and learning that this chapter therefore looks into the teacher education programmes that mathematics teachers in the four East African countries[1] of Kenya, Tanzania, Uganda and Rwanda receive before they embark on full-fledged teaching. It also looks into what professional development opportunities mathematics teachers have once they are practising the teaching profession. Both the initial teacher programmes and in-service teacher professional development programmes in mathematics teacher education in the region will be looked into from a comparative viewpoint, highlighting differences and similarities where applicable in mathematics teacher education curricula, student-teachers' entry qualifications into mathematics teaching, mathematics tutor qualifications and critically analyzing the implications of these differences or similarities to the regional move towards harmonized mathematics teacher education curricula and the overall quality of mathematics teaching and learning in the region. The chapter will conclude by making recommendations for policy and practice on the basis of the implications outlined above.

Pathways to Becoming a Teacher in East Africa

According to Thuranira (2010), teachers in Kenya are trained in teachers colleges and Universities. Holders of the Kenya Certificate of Secondary Education (KCSE), which is normally obtained by those who have studied in secondary schools for four years and successfully passed the national examination, are admitted to teachers colleges to pursue a two-year teaching course. Upon successful completion of this two-year course, graduates of these programmes are posted to teach in primary schools.

Secondary school teachers are trained through two channels: on the one hand, holders of the Kenya Certificate of Secondary Education (KCSE) are admitted to teachers' colleges earmarked to offer diploma programmes in secondary school teaching for a duration of three years. On the other hand KCSE holders are admitted to University undergraduate programmes where they major in education plus two teaching subjects for three years.

Tanzania's teaching force is recruited in teachers colleges and Universities. According to the Tanzania Institute of Education (2009), graduates of secondary schools are admitted to teachers' colleges to train to become Grade A teachers who then ultimately, upon successful completion of their two-year teaching

[1]The discussion will not include Burundi which is also part of the East African Community.

course, are posted in primary schools. There are also some teachers' colleges, less in number compared to those offering Grade A programmes, which are earmarked to offer teacher education for two years to prospective secondary school teachers at diploma level. Holders of teaching diplomas can only teach in forms one to four (i.e. Ordinary level, popularly referred to as O-level). Teachers who teach Advanced level classes (popularly referred to as A-level) are normally University graduates, who would normally have majored in education and two teaching subjects over a period of three years. Until quite recently, undergraduate programmes for those majoring in education had a four-year duration.

In Uganda, according to the Teachers' Initiative in Sub-Saharan Africa (TISSA 2013) report, there are two types of teachers' colleges: There are those that draw their student teachers from the lower secondary (i.e. O-level), who train to become primary school teachers. Then there are those that are known as National Teachers' Colleges, which draw student-teachers from the upper secondary schools to train to become secondary school teachers (both lower and upper). Universities also train undergraduate teachers who end up teaching in secondary schools.

In Rwanda, according to the World Bank (2011), graduates of the lower secondary school cycle are admitted to Primary Teacher Training Colleges (PTTCs) and upon successful completion of a three-year teacher training programme, they are posted to teach in primary schools. Lower secondary school teachers are trained at the University of Rwanda, College of Education (Kavumu or Rukara) for two years. The University of Rwanda, College of Education was until quite recently known as the Kigali Institute of Education. Successful completion of the training programme at the College of Education earns the student teachers a diploma in teaching. The College of Education of the University of Rwanda also undertakes teacher training at undergraduate level. The four-year undergraduate programme graduates are awarded a degree in education, which enable them to teach at both lower and upper secondary levels. They may also teach in teacher training colleges. The routes to becoming a teacher are summarized in the Table 5.1.

Considerable commonality in the four countries comes out in the analysis above. In the primary sector, a secondary school education, excluding the equivalent of A level, is sufficient for entry to teacher training college. It is noticeable that the total number of years starting at primary school varies, with Kenya requiring a minimum of 14, Tanzania and Uganda 13, whilst Rwanda requires only 12.

Similarly, all four countries require prospective secondary teachers to be educated up to completion of secondary school themselves, including A level. There is then a choice between training for a diploma or a degree, the degree entitling teachers to teach up to A level standard. Except for Rwanda, the training for diplomas is in colleges and for degrees in Universities. In Rwanda, which is a much smaller country than the others both in terms of land mass and population, all training is at the University of Rwanda College of Education, previously known as the Kigali Institute of Education.

In all four countries it takes a minimum of 16 years of schooling from the start of primary school to complete a degree, with 14 years for the diploma except

Table 5.1 Routes into initial teacher education

Country	Primary teacher training				Secondary teacher training			
	Type of institution	Entry qualification	No years to achieve entry from beginning of primary school	Duration in years	Type of institution	Entry qualification	No years to achieve entry from beginning of primary school	Duration in years
Kenya	Primary teacher college	Kenyan Certificate of Secondary Education	12	2	Teacher college/ University	Kenyan Certificate of Secondary Education	12	3 (diploma)/ 4 (degree)
Rwanda	Primary teacher training college	Completion lower secondary	9	3	College of Education	Completion upper secondary	12	2 (diploma)/ 4 (degree)
Tanzania	Teacher college	Certificate of Secondary Education	11	2	Teacher college/ University	Advanced Certificate of Education	13	2 (diploma)/ 3 (degree)
Uganda	Primary teacher training college	Ugandan Certificate of Education	11	2	National Teachers College/ University	Ugandan Advanced Certificate of Education	13	2/3

in Rwanda where it takes 13 years. Whilst this one year difference needs to be considered in the light of what actually happens in the schooling process, this is an issue which needs to be reconciled in moving towards harmonization.

Entry Criteria and Guidelines

As noted above, mathematics teachers for secondary schools are prepared either in teacher education colleges, which train them to the level of diploma in education, majoring in mathematics or in Schools of Education in Universities where they are trained to the level of degree holders. According to the Elimu Network (2013), only holders of C+ in mathematics in the Kenya Certificate of Secondary Education Examination (KCSEE) may be admitted to the Diploma Teacher Education course to study mathematics teaching. The candidates must also have attained an overall C+ mean score in his/her examination results in the KCSE (Kenya Certificate of Secondary Education). On the other hand, those proceeding to study education and mathematics teaching at the University level, would be required to have attained a plain C, although in case of admission to public Universities where competition is very high due to availability of many more facilities, the entry qualification requirement may be as high as a B.

In Tanzania, form six graduates (that is Advanced level graduates, those students who have had a total of 13 years of schooling) would be eligible for admission into diploma in education teachers colleges. They would however be required to have passed the Advanced Certificate of Secondary Education (ACSE) with principal passes or subsidiary passes in the two subjects that they intend to take as their teaching subjects. For a student teacher who intends to have mathematics as his/her teaching subject, he/she will be required to have a principal or subsidiary pass in mathematics in the ACSE examination.

According to the University of Dar es Salaam (2014), the minimum requirement for a form six graduate to qualify for admission to a University of Dar es Salaam undergraduate programme is principal level passes in appropriate subjects in the Advanced Certificate of Secondary Education Examinations (ACSEE) or equivalent with total points from three subjects not below 5 (for arts programmes) and 2 (for science-based programmes) based on the following grade to point conversion scale: A = 5; B = 4; C = 3; D = 2; E = 1; S = 0.5 and F = 0 point.[2] These main minimum requirements are stipulated by a number of many other Universities in the country. For a candidate who aspires to study mathematics as one of the teaching subjects, he/she would be required to have a principal pass in mathematics.

As for Uganda, a diploma in education for secondary school teaching is offered in what are known as National Teachers' Colleges (NTCs) and also at Universities. According to the TISSA (2013) report, completion of A-level with at least

[2]S stands for a subsidiary pass, A to E are principal passes while F stands for fail.

2 principal passes in art subjects or 1 in a science subject and two subsidiary passes for science subjects (which includes mathematics); plus O-level with at least 6 passes including English would be the minimum requirements for admission to any diploma in education course. Therefore, this means that a student teacher who intends to study mathematics at an NTC or University to earn a diploma in education for secondary school mathematics teaching, would require a principal pass in mathematics with two subsidiary passes in any other two science subjects, or a subsidiary pass in mathematics plus one principal and one subsidiary pass in any other science subject. With either option, six passes at O-level including a pass in English language and mathematics would still be mandatory.

The same report asserted that at University level, only those with two A-level principal passes in any of the arts, science and vocational subjects; or Grade V Teachers' Certificate would be eligible for undergraduate admission. For those intending to teach mathematics at secondary school level, they would need a principal pass in mathematics plus any other principal pass, or a Grade V Teachers' Certificate with two years of mathematics teaching experience.

To be admitted to a diploma in education programme in Rwanda, a candidate who intends to take up mathematics teaching upon completion must have two principal passes in science subjects one of which must necessarily be mathematics.[3]

Applicants for admission to undergraduate programmes who intend to study education majoring in mathematics and thereafter teach mathematics in Rwandan secondary schools must have two principal passes respectively in mathematics and one other science subject chosen from physics, chemistry, biology, accounting, and economics.[4] A minimum of a principal B in the given combination would be recommended for admission and an average of C in all subjects would be required. Table 5.2 summarises the entry requirements for these two different routes in the four countries.

Whereas it would appear that the admissions criteria are broadly comparable, there are discrepancies in what range of marks constitutes a given letter grade. For instance, an A in mathematics (and in all other subject for that matter) in Kenya would be in the range of 80–100 %, Tanzania would give an A to anything within the range of 75–100 %, while in Rwanda a candidate would need 85–100 % to earn an A. This would appear to mean that some candidates in one country would be eligible for admission into programmes (e.g. mathematics teacher education) while they would be denied admission with similar qualifications in another East African member country. This would have serious implications to the quality of education in the region. Some countries with robust grading systems would have better quality of education (assuming all other educational parameters are granted), which would inevitably afford them an advantage when it comes to accessing the labour market under the East African common market protocol.

---

[3]Information taken from the Rwanda Teachers' College, Kavumu campus website: http://rtckavumu.ac.rw/.

[4]Information taken from the University of Rwanda website: http://www.ur.ac.rw/.

Table 5.2 Entry criteria to the diploma and degree mathematics teacher programmes

	Secondary (diploma)	Secondary (degree)
Kenya	C+ in mathematics and C+ mean score in Kenya Certificate of Secondary Education	C+ in mathematics and C+ mean score in Kenya certificate of secondary education (in practice B required in public Universities)
Rwanda	Principal passes in mathematics and a science subject in A level	C in all subjects at A level, with B in mathematics
Tanzania	Principal or subsidiary pass in mathematics in A level	Principal pass grade A–E in A level
Uganda	Principal pass in mathematics plus 2 subsidiary passes in science plus 6 passes, including English at O level	Principal pass in mathematics and another subject OR Grade V Teachers' Certificate with 2 years' teaching experience

Mathematics Teacher Preparation: Structure and Curriculum

The Kenya's Ministry of Education and Ministry of Higher Education, Science and Technology in their sessional paper of 2012, outlined a policy framework on education and training. This stated that secondary teacher training at University level is provided using two models, the consecutive model and the concurrent model. In the former, the student teacher first earns a bachelor's degree (either in arts or science) majoring in their respective subjects. A nine-month postgraduate diploma programme focusing on pedagogy then follows this. In the latter, the subject content and pedagogy components are both dealt with at the same time and the student teacher earns a Bachelor of Education degree.

The diploma in education, leading to a mathematics teaching qualification in Kenyan secondary schools, is a three-year programme consisting of two strands. The first is the professional and support subjects and the second component is the academic one (Kenya Institute of Education 2008). The professional and support subjects include education, environmental education, PE, communication skills, entrepreneurship, ICT, general workshop practices, library and information studies, and guidance and counseling. These professional and support studies are studied by all along with two academic teaching subjects of their choice. For student teachers preparing to teach mathematics, the two academic subjects they would undertake to study would include mathematics as one of them.

The mathematics courses that are studied over the three years of the course consists of thirteen topics in the first year, three compulsory topics in their second year of study with two optional topics to be chosen from a total of four. Four topics are covered in year three bringing the total number of topics covered to twenty-two.

For the Tanzanian diploma a total of thirty-five mathematics topics are taught in the mathematics academic module over a span of two years TIE (2009), which consists of four units: foundations of mathematics; analysis of the essential mathematics curriculum materials; assessment in mathematics; and planning and preparation for teaching basic and additional mathematics.

Degree programmes in Tanzania specializing in mathematics take three years of study leading to a BSc with education degree, with the requirement that, in addition to studying mathematics, student teachers also major in education and an additional teaching subject.

In Rwanda the diploma in education programme consists of modules in education, entrepreneurship and communication skills with a total of 90 credits (900 hours) over the two years of the entire programme. Student teachers on the programme are also expected to do personal development planning courses, although these do not count towards their total number of credits. Student teachers who are trained to teach mathematics at lower secondary have to do 40 credits (400 hours) of modules in mathematics over and above the 900 hours of education, entrepreneurship, and communication skills (plus the non-credit personal development planning courses). Mathematics is always offered in conjunction with one other science subject. The College of Education of the University of Rwanda offers undergraduate programmes.

Uganda's degree programmes in education are offered in a number of Universities all over the country. However, Kyambogo University's Faculty of Education is renowned for its many undergraduate and post-graduate programmes in education and its role in the advancement of teacher education is widely acclaimed in the country (Ngobi, Otaala, Maani, & Bakaira 2011).

The diploma in education in science (including mathematics) is offered at four NTCs, namely Kabale, Kaliro, Muni, and Unyama. The programme runs over a span of two years. No information was available on the Internet that would have given an indication of how many hours are devoted to mathematics content and how many hours to pedagogy.

Of particular interest in the intended curricula for the four countries is the time durations that the diploma programmes in the different countries provide for the practicum component during the entire course which are summarized in Table 5.3.

The diploma programme in Kenya is 3 years rather than 2 in the other countries, but adding total teaching time since the beginning of primary school it is still rather more than Uganda's and Rwanda's which are broadly similar. It is noticeably Tanzania, which includes a very short amount of time for the practicum, from which one may well consider that the programme is largely theoretical. It is worth

Table 5.3 Intended curriculum for diploma in education

	Topics: academic	Topics: pedagogy	Teaching practice
Kenya	22 topics 618 h + 88 optional	8 topics 120 h	2 terms 26–28 weeks
Rwanda	900 h	400 h	1 term 10–13 weeks
Tanzania	12 topics 130 h	5 topics 128 h	4–5 weeks
Uganda			16 weeks

noting further that the practicum component is offered as a stand-alone and not as a continuing part and parcel of the theoretical components of the teacher education programmes. Student teachers go for practicum after they are done with their theoretical components of the programme. Without the opportunity to reflect on their practical experiences, it is difficult to see how student teachers are encouraged to see the connection between what is covered in theoretical sessions and the reality in the classroom, with serious implications to their efficacy upon completion of the programme.

This raises an issue that efforts towards educational systems harmonization in the region ought to focus on. In a bid to ensure comparability or compatibility of the curriculum and in particular of the practicum component, a long-enough duration should be allotted to practicum in the four countries in order to see to it that student-teachers get adequate hands-on experience in the field to be able to effectively function as professionals. Quality of mathematics teacher education in the East African countries will not be strengthened without the practicum issue being regulated.

Profile of the Tutors in Pre-service Teacher Education Colleges and Higher Educational Institutions

The quality of any teacher education programme depends, among many other factors, on the quality of the tutors who facilitate the programmes. This section looks at the profile of the tutors in pre-service teacher education programmes and lecturers in higher educational institutions that offer initial teacher preparation in mathematics teaching in terms of the requisite qualifications in ideal settings. This discussion is restricted to Tanzania and Uganda through unavailability of information about Kenya and Rwanda.

In Tanzania, it is stated policy that teachers' college tutors have to have a University degree, and the Teacher Education Development Programme has set for itself the target to recruit adequate quality teacher educators by ensuring that teaching in teachers colleges was both competitive and attractive to teachers with degree qualifications alongside upgrading all non-degree holders to degree qualification (The United Republic of Tanzania Basic Education Development Committee 2008). The same document goes on to highlight the fact that tutors in teachers' colleges were to be reflective, equipped with inquiry skills, and skilled in the use of ICT for teaching and learning. This is contrary to the author's experience as things stand at the moment, that many TC tutors do not have a degree.

At higher learning institutions it was observed that teacher educators were expected to be holders of doctoral degrees in educational studies (The United Republic of Tanzania Basic Education Development Committee 2008).

In Uganda, the national teachers' college lecturers are required to have a minimum of a bachelor's degree in education. As for Tanzania, no experience is required, and therefore a fresh graduate directly from education undergraduate programmes may straight away be posted to an NTC (The Republic of Uganda Education Service Commission 2011).

There are issues that arise from the above short discussion of profiles of tutors. On the one hand, permitting teaching personnel with qualifications less than the stipulated ones to work as tutors in teachers' colleges, as in the case of Tanzania, raises issues of quality. Whilst particular individuals may be excellent, it is reasonable to suppose that, in general, to allow holders of less than the prescribed qualification to operate is to compromise the quality of teacher education programmes. This ought to be one of the matters to be looked into as the East African block works towards harmonization of education systems. On the other hand tutors who have had no experience in teaching can, it is reasonable to assume, only be of very limited help in assisting others to learn how to teach. The practical experience that tutors with teaching experience bring to the teacher college is of crucial importance in providing examples of how teaching theories link to teaching practice.

Continuing Professional Development in Mathematics

Mizell (2010), extolling the need for professional development, posited that initial teacher programmes were not able to provide the 'extensive range of learning experiences necessary for graduates to become effective public school educators' and goes on to say that 'educators who do not experience effective professional development do not improve their skills, and students learning suffers'. This underlines the importance of teachers' professional development programmes in ensuring high quality education systems.

Teachers' continuing professional development generally speaking and in mathematics teachers' professional development in particular have been commended in all the four East African countries. This section dwells on the opportunities of continuing professional development that are available to mathematics teachers in the region.

Secondary School Teachers' Professional Development

There are a number of on-going in-service professional development programmes in science and mathematics with the aim of improving teachers' practice and subsequently, learners' performance in mathematics and science.

The Kenyan government has been working in collaboration with the Japanese government through the Japanese International Cooperation Agency (JICA) to improve the teaching and learning of mathematics and science since 1998, when it had become clear that secondary school teachers did not have opportunities to grow professionally and that the quality of education had immensely deteriorated (CEMASTEA 2013). The two governments jointly launched the first phase of the

Strengthening of Mathematics and Science in Secondary Education (SMASSE) programme in 1998 which lasted till 2003. Although initially the project was piloted in only nine districts, it was scaled up to cover the entire county in the second phase.

In Tanzania, JICA has been involved, in collaboration with the Ministry of Education and Vocational Training (MoEVT), in working with teachers to improve the mathematics and science teachers' skills in facilitating students' learning. There were national and regional groups of facilitators who worked with teachers in INSET centers all over the country and also at school level. Phase one of the project started in July, 2009 and it had a positive impact on teachers' classroom practice. More improvement is expected to be made in the on-going phase two of the project (Japan International Cooperation Agency 2014).[5]

In Rwanda, a project similar to the one that was embarked on in Kenya called SMASSE (Rwanda) project is on going and essentially aimed at strengthening the teaching of science and mathematics. In collaboration with the Rwandan Ministry of Education, JICA conducted a national situational analysis survey, which revealed, among many other problems, the challenge of teacher-centered pedagogy. Subsequently, INSET centers were established through out the country, one of which was earmarked as the national INSET Centre where core trainers trained about ninety national trainers. The latter were then deployed to work with secondary school mathematics and science teachers so as to help transform their teaching practice towards learner-centered pedagogy (Japan International Cooperation Agency 2014).

The SMASSE (Rwanda) project was aware that working with teachers alone would not produce excellent results if the other key players in teachers' working environment were not involved. For instance, curriculum development center personnel or the Teacher Service Commission personnel are so crucial to the teachers' working environment that they also needed to be sensitized to the core values of the project in which the teachers were participating. The SMASSE (Rwanda) project organized workshops to keep the Curriculum Development Center and the Teacher Service Commission personnel posted about the project's mission and vision.

Whereas the JICA project in both Kenya and Rwanda were named SMASSE, a similar JICA project in Uganda was called SESEMAT. SESEMAT, under the Ministry of Education and Sports (MoES) in collaboration with JICA, started its activities in Uganda in 2005. The project is currently in its third phase that started in 2013/2014 and is expected to end in 2017. The third phase was preceded by phases one and two both of which were characterized by the spirit of assisting mathematics and science teachers acquire practical knowledge and skills in order for them to improve their teaching practice. The INSET training had two strands: the teaching methodology strand and the mathematics/science content strand. The teaching methodology strand was guided by the ALEI/PIEI (Activity/Experiment,

[5]See the JICA website: http://www.jica.go.jp/english/.

Learner-centred, Encouragement, Improvisation, Plan, Implementation, Evaluation and Improvement) approach (SESEMAT 2014).

Although it is encouraging to note that there are mathematics teachers' professional development programmes in the region, it is still disheartening to find that each country has its own professional development programme running more or less independently of the other East African member countries. In this era when the East African community member countries are striving towards harmonizing their educational systems, it would have been convenient if more and more mathematics teachers' professional programmes were run as an East African enterprise. This would then have been a gesture in the direction of equally distributed improvement of the quality of mathematics education in the long run, which in turn would have ensured comparable quality of educational outcomes in the member states.

Tutors' Professional Development

The United Republic of Tanzania Basic Education Development Committee (2008) acknowledged that:

> In-service training for teachers, tutors and education managers is not adequately provided. It involves a limited number of education personnel. The provision of in-service training is currently neither regular nor continuous. To improve the capacity of delivery in the education system there is a need to ensure well resourced and continuous in-service training at all levels. (p. 11)

Within this overall bleak picture, very little is known about any professional development programmes that specifically target college tutors. There was a brief, undetailed mention by Chambulila (2008) of the existence of Tutors' Education Programme, which was aimed at developing tutors professionally.

In Uganda, it was categorically stated that, except for Makerere University, there was no express policy on staff development (Liang 2004).

> Staff professional development is not now part of the normal operations within any HE institution (except at Makerere University), and is treated as an additional expenditure that needs special funding. In part, this reflects the rigidity of the current funding regime since funds are allocated for specific purposes for most of the tertiary institutions. (p. 48)

The National Teachers' Colleges (NTCs), being part of tertiary or higher institutions of learning, are covered by this absence of laid down procedures to develop college tutors (called lecturers in Uganda).

The Rwandan teachers' management policy (Republic of Rwanda Ministry of Education 2007) very briefly talks about re-organizing teacher training in its new regulatory framework. The government of Rwanda commits herself to providing necessary inputs to support initial teacher training and the National Professional Development programme.

The situation in so far as tutors' professional development is concerned is a matter of concern, as it appears that it is not something that is taken with much seriousness. While it is important to continually professionally develop mathematics teachers, it is equally true that tutors, who interact with prospective teachers in initial teacher preparation, should also have opportunities to professionally develop themselves.

Conclusion

With free movement of services, labour and goods within the East African Community member countries under the EA common market protocol, what impact would stark differences in some areas of teacher education have on the quality of the services (product of education systems) provided by each of the East African states within the larger EA community context? For instance, it has been noted in the body of this chapter that the differences in the number of years spent on studying school mathematics may impact the quality of education in different member states. A harmonization of the teacher education systems is therefore imperative to allow for smoother movement of the envisaged services. Quality of teacher education programmes in each of the four countries in particular needs to be aligned one to another in such a way that graduates in each country would have more or less the same professional competencies and skills upon completion of their initial teacher education programmes in mathematics. This would be the surest way to guarantee quality services crossing the border from any of the member countries to any other.

Issues of similar grading systems, curriculum compatibility and comparability, alignment of professional development of both teachers and tutors, structure of practicum will all need to be addressed so that the principles governing these elements will be regularized and aligned to the best level possible without compromising the unique identity of the educational systems in all the four East African community member countries.

It is reassuring to note that the East African member states in fact started working towards harmonization of the entire education sector in 1998[6] (East Africa Community website). The crucial issues raised in this chapter will help to shape policy, which will in turn assist in shaping mathematics teacher education practice in the region towards a more efficient mathematics teaching and learning environment. This, as it was pointed out in the introduction, is the most pertinent precursor to the achievement of the country's respective development visions. The harmonization process therefore needs to be undertaken with all seriousness and urgency it deserves.

[6]See the East African Community website: http://www.eac.int/.

References

CEMASTEA. (2013). *Kenya SMASSE/SMASEE projects (1998–2013)*. Nairobi: Centre for Mathematics, Science and Technology In Africa.

Chambulila, C. D. (2008). *Teacher educators' professional development: the experience from tutors' education programme (TEP) in Tanzania*. Retrieved on 20th April 2015 from http://www.eera-ecer.de/ecer-programmes/conference/1/contribution/1119/

Constantine, J., Player, D., Silva, T., Hallgren, K., Grider, M., & Deke, J. (2009). *An evaluation of teachers trained through different routes to certification: Final report*. Jessup: Education Publications Center, U.S. Department of Education.

Darling-Hammond, L., Holtzman, D. J., Gatlin, S. J., & Vasquez Heilig, J. (2005). Does teacher preparation matter? Evidence about teacher certification, teach for America and teacher effectiveness. *Education Policy Analysis Archives, 13*(42), 1–51.

Government of the Republic of Kenya. (2007). *Kenya vision 2030*. Retrieved on 12th February 2015 from http://www.vision2030.go.ke/cms/vds/Popular_Version.pdf

Government of the Republic of Kenya. (2012). *A policy framework for education and training*. Nairobi: Ministry of Education.

Government of Uganda. (2013). *Uganda vision 2040*. Retrived on 20th April 2015 from http://gov.ug/content/uganda-vision-2040

Institute of Education. (2014). *History of teacher training in the UK: Institute of Education archives subject guide number 12*. London: Institute of Education, University of London.

Japan International Cooperation Agency. (2014). *In-service education and training for secondary school science and mathematics teachers phase II*. Retrieved on 20th April 2015 from http://www.jica.go.jp/tanzania/english/activities/education_01.html

Japan International Cooperation Agency. (2014). *Strengthening mathematics and science in secondary education (SMASSE Rwanda) project*. Retrieved on 20th April 2015 from http://www.jica.go.jp/rwanda/english/activities/activity01.html

KIE. (2008). *Diploma teacher education syllabus*. Nairobi: Kenya Institute of Education.

Liang, X. (2004). *Uganda tertiary education sector report: Africa region human development working paper series*. Washington, D.C: World Bank.

Lortie, D. C. (1975). *Schoolteacher: A sociological study*. Chicago: University of Chicago Press.

Mizell, H. (2010). *Why professional development matters*. Oxford: Learning Forward.

Moses, J. (2009). *The grass is greener.....a comparison of two distinct pathways to Qualified Teacher Status with particular reference to the expectations and experiences of secondary physical education trainees* (EdD). Durham: University of Durham.

Ngobi, D. H., Otaala, J., Maani, J., & Bakaira, G. (2011). The role of universities in teacher education and professional development: Kyambogo University, a case study. *CICE Series, 4*(2), 167–175.

Republic of Rwanda Ministry of Education. (2007). *Teacher development and management policy in Rwanda*. Kigali: Ministry of Education.

Republic of Rwanda, Ministry of Finance and Economic Planning. (2000). *Rwanda vision 2020*. Kigali: Ministry of Finance and Economic Planning.

SCETT. (2011). *In defence of teacher education: a response to the Coalition Government's White Paper for Schools (November 2010)*. Derby: Standing Committee for the Education and Training of Teachers.

SESEMAT. (2014). *Newsletter issue 1*. Kampala: JICA.

Teachers' Initiative in Sub-Saharan Africa. (2013). *Teacher issues in Uganda: A diagnosis for a shared vision on issues and the designing of a feasible, indigenous and effective teachers' policy*. Kampala: Ministry of Education and Sports.

The Elimu Network. (2013). *Regulations for the diploma teacher education course*. Retrieved April 4, 2014, from http://kenya.elimu.net/Diploma_Teacher_Education/DTE_Regulations.htm

The Republic of Uganda Education Service Commission. (2011). *Scheme of service for teaching personnel in teacher education institutions*. Kampala: Education Service Commission.

The United Republic of Tanzania. (1999). *Tanzania Development Vision 2025*. Retrieved on 20th April 2015 from http://www.mipango.go.tz/index.php?option=com_docman&Itemid=107

The United Republic of Tanzania Basic Education Development Committee. (2008). *Education sector development programme: the teacher development and management strategy (2008–2013)*. Retrieved on 8th April 2014 from http://www.ed-dpg.or.tz/pdf/TE/Teacher%20Education%20and%20Management%20Strategy-TDMS%20-final%202009.pdf

Thuranira, T. S. (2010). *Perspectives on the teaching profession in Kenya*. Unpublished PhD thesis, University of Maryland.

TIE. (2009). *Curriculum for diploma in teacher education programmes in Tanzania*. Dar es Salaam: Tanzania Institute for Education.

University of Dar es Salaam. (2014). *Entry requirements*. Retrieved April 8, 2014, from https://udsm.ac.tz/?q=undergraduate-entry-requirements

Wolhuter, C. C. (2006). Teacher training in South Africa: Past, present and future. *Education Research and Perspectives, 33*(2), 124–139.

World Bank. (2011). *Rwanda education country status report: Toward quality enhancement and achievement of universal nine year basic education*. Washington, D.C.: World Bank.

Chapter 6
ICT Integration in Mathematics Teaching and Learning: Insights from East Africa

Alphonse Uworwabayeho

Abstract In order to prepare young people for participation in the technological and knowledge driven society, policy makers in the East African countries are emphasizing the use of ICT to transform the economy from subsistence agriculture to a knowledge-based economy. With this perspective, teacher training in ICTs should not just be about using new technologies but also about why and when to use them in transforming teaching and learning practices. For example, objectives of the Rwandan ICT in Education Policy Statement include developing teachers' capacity and capability in and through ICT at all school levels. It is against this objective, the present chapter aims at analyzing what level of use of ICT is currently supporting the teaching and learning of mathematics in basic education in Eastern Africa region with illustrating examples from Rwanda. Whilst ICT helps teachers for demonstrations and presentations of their material, group work enhances the learning collaboration; thus breaking away from the practice of talk and chalk that is mostly teacher-centred to a more learner-centred teaching approach. The case study of Rwanda raises important questions for harmonization of mathematics education in the East Africa region.

Introductory Background

The last few decades have seen huge developments in technology. Whereas only 50 years ago computers filled entire rooms, there are now laptops, tablets, smart phones and many other devices with increasingly sophisticated software, becoming ever more powerful, cheaper, smaller and better connected, more and more visible in many walks of life including education.

A. Uworwabayeho (✉)
School of Education, University of Rwanda-College of Education,
Po Box 5039, Kigali, Rwanda
e-mail: rwabayeho@yahoo.fr

A. Halai and G. Tennant (eds.), *Mathematics Education in East Africa*,
https://doi.org/10.1007/978-3-319-27258-0_6

Broadly speaking the role of ICT within education is twofold. One is to prepare youngsters with the skills and attitudes needed to operate within a technological world. The other is to enhance the learning process, with ICT offering opportunities not easily available in other ways.

This chapter considers some of the opportunities and challenges faced in looking to introduce ICT into the classroom, considering the issues firstly generally, in the developing world and specifically in East Africa. Some case studies from Rwanda in which ICT is used in the mathematics classroom are then considered, showing how ICT can be used meaningfully in a low income context, drawing implications as to how this key area of education can be developed further.

How Does ICT Get Used in the Classroom?

In principle there is large number of ways in which ICT can be deployed in the classroom. Considering first laptop or desktop computers, methods include:

- lecture by the teacher supported by presentation software e.g. PowerPoint;
- demonstration by the teacher using dynamic geometry software e.g. Geometer's Sketchpad;
- allowing children to use a small number of computers (maybe one) individually in small groups whilst the rest of the group is working on something else;
- small number of children using the computer whilst the teacher guides the whole class in the work.

Further devices, e.g. cameras, smart phones and tablets, offer further possibilities, e.g. the recording of science experiments and field trips, real time 'voting' and many others.

Alongside an examination of what is available in terms of hardware and software, it is essential also to consider the implications for pedagogy. As becomes clear in a literature review prepared in a UK context (Cox, Webb, Abbott, Blakeley, Beauchamp, & Rhodes 2003) the mere availability and indeed use of ICT does not mean that any changes are being made to the underlying dynamic of the classroom. Indeed, one might well consider that PowerPoint or equivalent software is ideal for supporting a very teacher-centred didactic form of pedagogy. ICT can, however, present opportunities for supporting learner-centred pedagogy, collaborative investigative work, if teachers are confident ICT users and have strong subject and pedagogic knowledge, are well supported in the work that they do.

ICT in the Developing World

Specifically within East Africa the use of ICT in classrooms is very much on the agenda, with countries' national development vision plans centring around creating "a prosperous knowledge-based economy" based on ICTs and education playing a key role (e.g. Rwanda Vision 2020 and Tanzania Vision 2025, see MINECOFIN 2003 and URTPC 2003 respectively). Various Kenyan policies on ICT have been motivated by improving the livelihoods of Kenyans by ensuring the availability of accessible, efficient, reliable and affordable ICT services and the development of e-learning based materials, including the adoption of a one laptop per child policy, albeit that evidence from other countries as to the efficacy of this policy is mixed (in a Peruvian context see Trucano 2012). The Ugandan policy is based on the premise that ICT use is a key skill required for a rapidly increasing range of jobs, and developing good ICT skills in young people can help them find employment (Hennessy, Onguko, Harrison, Ang'ondi, Namalefe, et al. 2010).

For East Africa as a whole, there has been remarkable growth in using Information and Communications Technology (ICT) in education within the past decade; and most of the expansion in ICT tools has been in computers. On one hand, this investment is motivated by the vision that ICT in schools will support young people to develop knowledge and skills that enable them to participate fully in society. On the other hand, providing schools with ICT equipment and resources is only a starting point in a process of empowering young people to use ICTs to transform their learning and their lives. The investment will only impact in such a way to both transform learning and educate young people to contribute to building the economy if young people are allowed access to hands-on activities with ICT, and this requires teachers to develop a more student-centred approach (Uworwabayeho, Rubagiza, Olivero, & Sutherland 2013). In the same line, EdQual research findings (EdQual 2014) indicate that teacher professional development is a key in promoting student-centred approach to use ICT in the classroom, with Hennessy et al. (2010) indicating that there is considerable work to do in ensuring that teachers feel comfortable with computers and view them as the useful, enabling tool they can be.

Initiatives in ICT Implementation in East African Schools

A review of literature assessing the development of the use of ICT to enhance teaching and learning in East African schools identified a number of important physical, cultural, socioeconomic and pedagogical factors hindering the use of ICT by teachers and students in sub-Saharan Africa, particularly in rural schools (Hennessy et al. 2010). These include lack of electricity and frequent power outages, poor technology infrastructure, overcrowded computer labs and low bandwidth, high costs of (mainly satellite) internet connectivity, software licenses and equipment maintenance, insufficient and inappropriate software. If, in addition, the teaching style used is predominantly didactic (see Halai in this volume) then the discussion above would indicate that the effectiveness of the implementation of ICT will be limited.

On the other hand, all East African countries have found external funding in supporting initiatives towards ICT in education. To name a few of these initiatives, Kenya, Uganda and Rwanda benefited from the New Africa's Development (NEPAD) e-school initiatives in 2006. This was a continent-wide initiative piloted in 20 countries with the aim of building ICT capacity and raising the quality of education in African schools. During the period 2006–2010 the NEPAD e-School Initiative was represented in those 20 African countries among the 3 mentioned above each with 6 schools. In this way each of six Rwandan secondary schools was provided with 20 computers, Internet connectivity, projectors, laptops and interactive whiteboards, ICT skills training packages, subject related materials and support (troubleshooting and coordination).

SchoolNet in Uganda was a national network of professional educators project whose vision was to transform the Uganda educational system from an industrial model (learning by assimilation) to a knowledge-based model to prepare the youth of Uganda to effectively enter a global economy based on knowledge, information and technology. Activities included equipping schools with PCs and developing localised content that maps directly to the national curriculum. Through donated PCs from Microsoft and other entities, the country was able to equip 100 schools with approximately 10 computers each (Hennessy et al. 2010).

The Ministry of Education in Rwanda has overseen the expansion of ICT infrastructure through a number of initiatives in primary and secondary schools. Starting in the late 1990s, there was a rollout of providing computers to primary schools through the collaboration of the Ministry of Education and World Links, a USA-based non-governmental organisation. Two desktop computers were provided to each of 98 primary schools that had an electricity supply and one laptop plus solar supply to each of 1018 primary schools without electricity. The government through the Ministry of Education continued to supply secondary schools with computers and by the end of 2006, 400 out of the 500, private and public secondary schools in Rwanda at that time had received 10 computers each (Rubagiza, Were, & Sutherland 2011). Another element that led to opportunities for integration of ICT in the teaching and learning of mathematics is given from huge potential local and international development partnership with private companies and Rwandan government. In this line, existing national, cross-border, and submarine communication network enabled the country be covered by an optical fiber cable. To date, the remaining task is for each school/learning institution to connect to this main cable.

The Rwandan Economic Development and Poverty Reduction Strategy (EDPRS) presents a vision that involves the creation of a cadre of students with well-developed ICT skills, which may be useful for future studies and integration of Rwandan businesses into the global economy by 2012 (MINECOFIN 2007). To achieve this, the policy continues to state that teachers must acquire competencies and confidence to integrate ICT into their teaching routine, and develop innovative classroom environments using ICT tools that motivate and improve learning/teaching process, helping learners to think critically and creatively and to communicate their thinking. In particular, the education sector policy emphasizes that (science and) mathematics teaching and ICT shall be at the heart of the education system (MINEDUC 2010).

Although there exists a clear ICT policy in education, its implementation faces challenges that include technical (capable human resources) but more importantly collaboration between different systems in charge of education in Rwanda

(Hooker 2009). This is a common challenge within the region. For example analysis on ICT in education in Tanzania reveals that a lack of coordination of ICT in education activities, limited information sharing, ineffective organizational structures at the various education management levels are part of constraints towards the integration of ICT in the teaching and learning (Swarts and Wachira 2010). There is also evidence that the presence and use of ICT in schools in Rwanda is still minimal with respect to teaching and learning mathematics and multiple challenges are faced. These include inadequate resources, such as the low number of computers compared to expanding numbers of student populations, limited ICT skills among teachers, lack of electricity supply in some schools and problems of maintenance (Rubagiza et al. 2011). To date, computer to student ratio in secondary schools is 40:1 whilst only 16 % of primary schools use XO laptops.[1] As a consequence there is a tension between those teachers who use computers for teaching and learning ICT skills and those who want to use ICT for the teaching and learning of mathematics. The latter group is increasingly finding it difficult to access the computers, since the use of ICT to teach other subjects is not yet seen as a priority, given the relatively scarce ICT resources in schools (EdQual 2014). A paradox is revealed by mathematics teachers who may find support in utilizing the existing ICT resources but have little time for planning. A lesson using ICT is seen as requiring a large amount of preparation not only in terms of the logistical issues of where and when it might take place but also the collection of resources and developing appropriate activities. In short, provision of ICT in schools is only the first step towards its embedding in teaching and learning (EdQual 2014). Swarts and Wachira (2010) conclude that the strong commitment of regional governments to using ICT for educational changes is the only opportunity for integrating ICT in the teaching and learning mathematics. But on the other hand, EdQual research findings reveal that with relatively small amounts of external support through workshops teachers become confident in exploiting the existing ICT tools to develop innovative mathematics classroom environment. The following section discusses the EdQual use of participatory action-oriented inquiry for teacher professional development on integration of ICT in the teaching of mathematics.

Teachers Professional Development Model Towards Integration of ICT in Teaching Mathematics: A Rwandan Case-Study

This section presents an example of an intervention on teacher training in using ICT in the teaching and learning of mathematics. The section starts with outlining the collaboration between academics and mathematics teachers, then classroom

[1]The XO laptop was designed alongside the 'One laptop per child' initiative to be 'durable, functional, energy-efficient, responsive, and fun' (see http://one.laptop.org/about/hardware).

case studies. One case study is based on the use of Excel while another uses a particular software for geometry, namely Geometer's Sketchpad. The intervention was initiated by 'EdQual-ICT in Basic Education' one component of the large scale research project namely 'Research Programme Consortium on Implementing Education Quality in Low Income Countries (EdQual)' funded by DfID (UK Department for International Development). The EdQual-ICT in Basic Education project was grounded within the Rwandan National Development Vision that the country will achieve middle income status by 2020. The Vision 2020 plan centres around creating "a prosperous knowledge-based economy" based on information and communication technologies, thus hoping to make Rwanda the ICT hub of Africa (Uworwabayeho et al. 2013). The EdQual-ICT in Basic Education project adopted the "evolutionary model for teachers' professional development" for research design in Rwanda for the five years between 2005 and 2010. This process built on the work of the InterActive Education Project in UK (Sutherland, Robertson, & John 2008) in the sense that through a series of teacher workshops, participant mathematics teachers from 12 schools (primary and lower secondary schools) would work together to develop their own skills and to plan activities to be carried out in their own schools. The workshops, which were generally scheduled at the end of every school year, usually lasted 4–5 days. They were used to provide time to develop knowledge and skills, especially with regard to ICT use in the teaching and learning of mathematics, but also other more general pedagogical and ICT skills. Sessions were initially led by local project team members from University of Rwanda-College of Education (former Kigali Institute of Education) and international partners (from Chile and United Kingdom) who were able to share ideas and present examples of ICT use in the classroom. As teachers became involved in research in their own schools, they were able to contribute to the workshops themselves, presenting work they had done and suggesting ways forward. The role of the academics became to support these processes at the workshops themselves and to monitor the implementation of planned activities through visits to the schools in the periods between workshops. In addition to sharing experiences and reflecting on the process of the research; and future plans and schedules, teachers' workshops also offered a time to develop some skills, especially with regard to ICT use in the teaching and learning of mathematics and also other pedagogical skills that relate to a learner-centred approach (Uworwabayeho et al. 2013). Therefore, the collaborative professional development model focused on teachers developing innovative classroom scenarios through active 'hands-on' experimentation with ICTs and reflection and discussion with other teachers. Video data was invaluable in both sharing practice and for analysing teaching and learning. Initial classroom observations revealed that the use of ICT could not provide change in teaching approach that remained mainly teacher-centred but as the time evolved, teachers became more and more creative in organizing interactions that embed a learner-centred teaching approach (Uworwabayeho 2012).

The collaborative approach used in the workshops extended to the in-school support provided by the University project team. School visits by researchers played an important role in maintaining the momentum of the project. The experience of having observations from someone other than an 'inspector' was a new one for many of the teachers. The model used in the project for school visits was at collegial/peer level with a focus on improvement rather than a hierarchical and judgmental model. In some cases, the experience made teachers more open to peer observations with colleagues in their own schools because they saw the value of this in helping others but also in reflecting on their own teaching. The following section illustrates some classroom scenarios to illustrate the impact of EdQual intervention.

Teaching and Learning of Mathematics: Stories from Classrooms

Initial suggestions for use of ICT to support teaching and learning in mathematics focused on the use of spreadsheets. Though spreadsheets were not purposely designed for teaching and learning mathematics, research in education showed that they can facilitate the introduction of algebraic concepts such as functions, variables, general formulas, parameters, equivalent expressions and symbolisation of numeric and geometric patterns, as well as basic statistical skills (Sutherland 2007). Although there is a large amount of software for sale, with limited funding available, the use of spreadsheets provided a good starting point to explore a number of areas for organizing and analyzing data. One example observed in school was using a spreadsheet to record prices of various commodities and foods and use a simple formula to calculate costs of different quantities of each. More complex use of formulae can support learning of algebra and statistics. Thus, a widely available software tool (Excel was a programme on all computers in the project schools) provided a wide range of opportunities to support mathematical learning.

Example 1: mathematics lesson in primary school using spreadsheets
The room is equipped with 2 computers and a group of 16 learners at a time are called into the computer room. Two mathematics teachers are leading the class. The exercise was written on the chalkboard and 2 learners enter data into the computers. An incomplete shopping list is given, with each item on the list showing with quantity, unit price and total price. The learners' task is to work out the missing information and finally the total amount paid for all items. Mathematics operations involved are: multiplication, division and addition. Learners are then invited to fill in the missing data using spreadsheets formula, e.g., '=D2/B2' for the unitary price of 'beans' or '=B5 * C5' for total price of 'sugar'. While only two learners are actually using the computers, others are following and can interact with one of the teachers.

Example 2: mathematics in secondary school using dynamic geometry
A second example comes from collaboration between the chapter author and two mathematics teachers using 'Geometers' Sketchpad' software (Uworwabayeho 2009). The figure below is an example of working of a group of three learners using the software to construct and compare the sums of two vectors. The lesson

took place on 17th February 2010 in the school computer lab. The school was a NEPAD e-school as described above. There were 28 students (in year 3, ordinary level) sitting in 12 groups in total, each comprising 2 or 3 students. The lesson was organized in such way that students started with working in small groups, then the teachers organized whole class discussions where each group representative presented group findings. Students were asked to work on the following activity:

Let $\vec{u}$ and $\vec{v}$ be vectors. Construct the vectors $\vec{u}+\vec{v}$ and $\vec{v}+\vec{u}$. What is the relationship between them? Give a conclusion.

The overall lesson topic was about properties: commutativity and associativity of the addition operation in the set of vectors. Working on questions prepared by the teacher, students had to construct the sum of vectors in dynamic geometry and 'discover' these properties. Students were introduced to the sum of two vectors using dynamic geometry during the previous lesson. They were also familiar with the concepts of commutativity and associativity of addition in number sets (integers and real numbers) from their previous academic years. Within the lesson, there was an absence of teacher-student interactions at whole classroom level; students put more time in experimenting with the dynamic geometry where they were asked to apply mathematical theory (projection) to construct sums of vectors. They also had to read feedback from the dynamic geometry to formulate mathematical knowledge. The teacher was moving around to offer assistance and could request students to explain the process which they went through in constructing the sum of vectors.

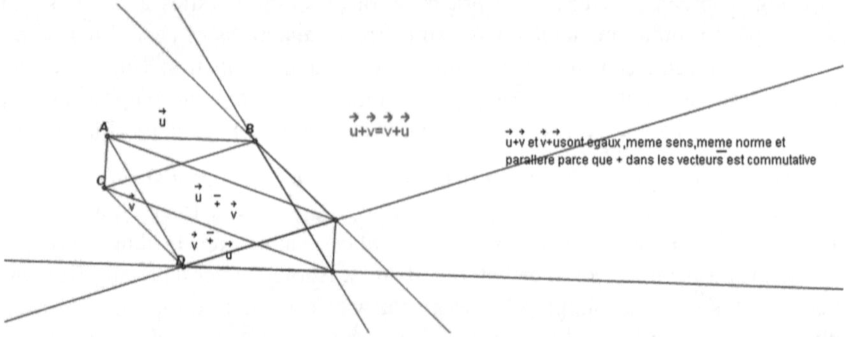

Learners' construction in Geometers' Sketchpad (17/2/2010)

This sort of software allows students to work through problems and test out their ideas while learning basic principles. Firstly, student need to know the commands to use for constructing the two vectors $\vec{u}$ and $\vec{v}$; then the mathematical procedure for constructing the sum of two vectors since the construction depends on how the students represent the two vectors, for example whether $\vec{u}$ and $\vec{v}$ have the same origin or not. In this case, dynamic geometry can facilitate or constrain students in developing the new mathematical knowledge (hidden in the rubric, *give a conclusion*). In the former case, dynamic geometry can provide positive feedback by facilitating visualization. In the latter case, the student fails in constructing vector sums (antagonist context). Secondly, the student must deal with the mathematical constraints of:

what is the relationship?; and what is the conclusion? For example students need to know under which conditions two vectors are said to be equal (cognitive context). Therefore the activity potentially allows students to think and speak. In addition, it was observed that manipulating drawings with the software kept students trying different possible answers to the question.

These classroom scenarios illustrate that, the teacher acted to increase the potential for action in a minimal way so that students had the opportunity to achieve as much as possible by themselves especially when they encountered difficulties with their tasks. Thus, the use of the dynamic geometry in this teaching seems to promote students' creative and critical thinking as well as conceptual learning.

Students' work in groups on computers encourages collaboration and cooperation. Whilst there were only a very few computers available for the number of students, therefore not all students were able to experience hands-on use of ICT, it was nevertheless the case that the model of collaboration undertaken enabled participant teachers to change their practices. It is argued that reflection on one's actions constitutes a pillar of his/her own professional development though not sufficient on itself (Day 1999). Despite these achievements, teacher participants pointed out that use of ICT is time consuming in terms of lesson preparation and hides difficulties in classroom management especially in large size classes. Moreover teachers argued that in order that ICT should be entirely integrated in the teaching and learning there is a need to organize classroom based support from a trainer, who regularly observes and discusses practice with the teachers and encourage teachers at the same school to develop their classroom practice as a team.

Conclusion: Towards Regional Harmonization

This section draws some conclusions on issues that need to be taken into consideration in order to integrate ICT in the teaching and learning of mathematics in East African community. These issues are related to the implementation of policies, physical infrastructures in schools and mathematics teachers' professional development towards use of ICT.

The vision that ICT will both transform learning in schools and educate young people to contribute to building the economy has been echoed in high and low income countries around the world. Within the East African region, there is a strong government awareness and commitment to ICT in education; and a number of initiatives have been implemented to provide pedagogical and subject support to teachers though this is in most cases limited to ICT skills without the integration in the teaching (Swarts and Wachira 2010). Moreover, at country level and to some extent at regional level, there is a need to build a strong coordination mechanism among ICT-related initiatives in schools. The emphasis should be using technological solutions that are suited to regional/local needs and conditions. Curriculum design, delivery and assessment should establish appropriate mechanisms for regulating the development and use of electronic content. Finally a mechanism of offering pre- and in-service training for mathematics teachers on a continuous basis to keep up to date with technological and pedagogical developments should be established. The implementation of this policy must be conducted via a partnership approach involving the community, private and public organizations, and development partners. It is indubitable that teachers are the key in promoting education quality that and implementing new education innovative. In this line, mathematics teachers' skills, expectations and usage of ICT are essential and need to be developed, through appropriate professional development.

Examples from Rwandan classrooms presented above show how ICT tools can be used as catalysts to provoke mathematics teachers to change their practice. However as it is elsewhere (Sutherland et al. 2008) there is nothing inherent in the ICT tool itself that 'causes' change, it is how the ICT tool is used that is important. And how the tool is used by teachers relates to professional development practices. The mathematics examples particularly show that it is possible to encourage the emergence of a student-centered model of mathematics and science teaching through employing a participatory action oriented-research based approach to teacher professional development. And within this approach, the use of ICT appears to be inextricably bound up with the process of change.

Teacher professional development networks should be promoted throughout the country to some extend in the region. Building on the model developed in the EdQual project in Rwanda, teachers should be supported to allow student-centered learning with ICT, exploit available technology including mobile technology through existing initiatives for example "One laptop per child project", understand how out-of-school use of ICT impacts on learning in school and develop positive strategies to address this, and contribute to quality education by communicating innovative classroom practices.

If we consider teaching as organizing the milieu in order to allow the student to independently reach the new mathematical knowledge, it would be plausible to suggest that there is a necessity of developing familiarity with technology in use if we want users to appropriate it. Further research findings (e.g. Laborde 2001) have shown the difference between 'new' and 'old' users in exploiting ICT tools for teaching and learning purposes. In particular, Laborde points out that the efficiency of using ICT in teaching and learning depends much more on the choice of mathematics activities which in turn depends on teacher's pedagogical skills as well as the familiarity with the tool. So, questions remain on what type of mathematical activities fits with existing ICT tools within a school. In addition to the tool in use, mathematical topics stated by the curriculum play a role in organizing the teaching and the learning process. Primordially, teachers need to respond to curriculum exigencies. Setting the classroom activities, the teachers need to take into consideration the students' prerequisite knowledge, which makes it difficult to create suitable situations where students organize their learning independently to the teacher's direction at a given period of teaching. Some topics might well fit within such kind of situations while others would not.

In most East African schools, unsupervised students do not have access to ICT tools (more specifically computers); but also mathematics teachers are limited in using them for teaching mathematics due to school constraints and curriculum constraints. These include few computers compared with the school population size; thus priority is determined by the national education policy that developed a curriculum for ICT literacy. As a consequence, mathematics teachers need to take into consideration students' unfamiliarity with ICT, therefore slowing down the progress in curriculum content against which teaching and learning are judged. Therefore, despite teachers' initiatives in exploiting ICT for effective teaching and learning, they are inhibited by environmental conditions. As suggested elsewhere (Uworwabayeho et al. 2013) what is needed for ICT to be integrated in education in a way that empowers young people's experiences and life, is a participatory approach to professional development, in which teachers are involved in formulating the research questions and participating in the design and development of teaching scenarios. In this way, whilst short workshops can be effective for providing teachers with techniques to use in integrating ICT in the teaching and learning of mathematics, a more collaborative professional development in schools is needed to transform teachers' practice in line with the demands of ICT and competencies based-curricula now being adopted across East African countries.

References

Cox, M., Webb, M., Abbott, C., Blakeley, B., Beauchamp, T., & Rhodes, V. (2003). *ICT and pedagogy: A review of the research literature*. London: DfES.

Day, C. (1999). *Developing teachers: The challenge of lifelong learning*. London: Falmer Press.

EdQual(2014). *EdQual, a research programme consortium on implementing education quality in low income countries: Final report 05-11*. Bristol, UK: EdQual.

Hennessy, S., Onguko, B., Harrison, D., Ang'ondi, K. E., Namalefe, S., et al. (2010). *Developing the use of information and communication technology to enhance teaching and learning in East African schools: Review of the literature*. Cambridge, UK: Faculty of Education.

Hooker, M. (2009). How can I encourage multi-stakeholder narrative and reflection on the use of ICT in teacher professional development programmes in Rwanda? *Educational Journal of Living Theories, 2*(3), 324–364.

Laborde, C. (2001). Integration of technology in the design of geometry tasks with Cabri-geometry. *International Journal of Computers for Mathematical Learning, 6*, 283–317.

MINECOFIN. (2003). *Vision 2020*. Kigali: Republic of Rwanda.

MINECOFIN (2007). *Economic development and poverty reduction strategy 2008–2012*. Kigali: Republic of Rwanda.

MINEDUC. (2010). *Education sector strategic plan 2010–2015*. Kigali: Republic of Rwanda.

Rubagiza, J., Were, E., & Sutherland, R. (2011). Introducing ICT into schools in Rwanda: Educational challenges and opportunities. *International Journal of Educational Development, 31*, 37–43.

Sutherland, R. (2007). *Teaching for learning mathematics*. Maidenhead, UK: Open University Press.

Sutherland, R., Robertson, S., & John, P. (2008). *Improving classroom learning with ICT*. London: Routledge.

Swarts, P. & Wachira, E. M. (2010). *Tanzania: ICT in education, situational analysis*. Retrieved December 11, 2014 from http://creativecommons.org/licenses/by-nc-sa/3.0/

Trucano, M. (2012). *Evaluating one laptop per child (OLPC) in Peru. 2/26/2015*. Retrieved February 21, 2015 from http://blogs.worldbank.org/edutech/olpc-peru2

URTPC. (2003). *Tanzania development vision 2025*. Dar es Salaam: United Republic of Tanzania Planning Commission.

Uworwabayeho, A. (2009). Teachers' innovative change within countrywide reform: a case study in Rwanda. *Journal of Mathematics Teacher Education, 12*(5), 315–324.

Uworwabayeho, A. (2012). *Investigating Rwandan mathematics teachers' shifts towards learner-centred approaches when using ICT: A participatory collaborative action-oriented inquiry*. Unpublished PhD thesis, University of Bristol, UK.

Uworwabayeho, A., Rubagiza, J., Olivero, F., & Sutherland, R. (2013). Participatory professional development: ICT and mathematics education in Rwanda. In L. Tikly & A. Barret (Eds.), *Education quality and social justice in the south: Challenges for policy, practice and research* (pp. 181–196). London: Routledge.

Correction to: Mathematics Education in East Africa

Correction to:
A. Halai and G. Tennant (eds.), *Mathematics Education in East Africa*, https://doi.org/10.1007/978-3-319-27258-0

The original version of the book was published in 2016 with exclusive rights reserved by the Publisher. As of December 2020 it has been changed to an open access publication: © The Editor(s) (if applicable) and The Author(s) 2016.
The copyright of the individual chapters has therefore changed to: © The Author(s) 2016
All chapters in the book are licensed under the terms of the Creative Commons Attribution-NonCommercial-NoDerivatives 4.0 International License.
Any third party material is under the same Creative Commons license as the book unless specified otherwise below.

The updated version of this book can be found at https://doi.org/10.1007/978-3-319-27258-0

A. Halai and G. Tennant (eds.), *Mathematics Education in East Africa*,
https://doi.org/10.1007/978-3-319-27258-0_7